Science in World History

T0144592

Science today is a truly global enterprise. This book is a comprehensive, thematic survey of the history of science from its roots in different cultures around the world through to the present day.

James Trefil traces how modern science spread from its roots in Western Europe to the worldwide activity it is today, exploring crucial milestones such as the Copernican revolution, the germ theory of disease, and the theory of relativity. In doing so, he also examines the enormous social and intellectual changes they initiated. Opening with a discussion of the key elements of modern scientific enterprise, the book goes on to explore the earliest scientific activities, moving through to Greece and Alexandria, science in the Muslim world, and then on to Isaac Newton, atomic theory and the major developments of the nineteenth century. After examining the most recent scientific activities across the world, the book concludes by identifying future directions for the field.

Suitable for introductory courses and students new to the subject, this concise and lively study reconsiders the history of science from the perspective of world and comparative history.

James Trefil is Clarence J Robinson Professor of Physics at George Mason University. Author of over 40 books and the recipient of numerous awards, he is renowned for his ability to explain science to non-scientists.

Themes in World History
Series editor: Peter N. Stearns

The *Themes in World History* series offers focused treatment of a range of human experiences and institutions in the world history context. The purpose is to provide serious, if brief, discussions of important topics as additions to textbook coverage and document collections. The treatments will allow students to probe particular facets of the human story in greater depth than textbook coverage allows, and to gain a fuller sense of historians' analytical methods and debates in the process. Each topic is handled over time – allowing discussions of changes and continuities. Each topic is assessed in terms of a range of different societies and religions – allowing comparisons of relevant similarities and differences. Each book in the series helps readers deal with world history in action, evaluating global contexts as they work through some of the key components of human society and human life.

Gender in World History
Peter N. Stearns

Consumerism in World History: The Global Transformation of Desire
Peter N. Stearns

Warfare in World History
Michael S. Neiberg

Disease and Medicine in World History
Sheldon Watts

Western Civilization in World History
Peter N. Stearns

The Indian Ocean in World History
Milo Kearney

Asian Democracy in World History
Alan T. Wood

Science in World History

James Trefil

Routledge
Taylor & Francis Group

LONDON AND NEW YORK

First published 2012
by Routledge
2 Park Square, Milton Park, Abingdon, Oxon OX14 4RN

Simultaneously published in the USA and Canada
by Routledge
711 Third Avenue, New York, NY 10017

Routledge is an imprint of the Taylor & Francis Group, an informa business

British Library Cataloguing in Publication Data
A catalogue record for this book is available from the British Library

Library of Congress Cataloging in Publication Data
Trefil, James S., 1938-
Science in world history / James Trefil.
p. cm. – (Themes in world history)
Summary: "Science today is a truly global enterprise. This book is a
comprehensive, thematic survey of the history of science from its roots in
different cultures around the world through to the present day. James Trefil
traces how modern science spread from its roots in Western Europe to the
worldwide activity it is today, exploring crucial milestones such as the
Copernican revolution, the germ theory of disease, and the theory of
relativity. In doing so, he also examines the enormous social and intellectual
changes they initiated. Opening with a discussion of the key elements of
modern scientific enterprise, the book goes on to explore the earliest
scientific activities, moving through to Greece and Alexandria, science in the
Muslim world, and then on to Isaac Newton, atomic theory and the major
developments of the nineteenth century. After examining the most recent
scientific activities across the world, the book concludes by identifying future
directions for the field. Suitable for introductory courses and students new to
the subject, this concise and lively study reconsiders the history of science
from the perspective of world and comparative history"– Provided by publisher.
Includes bibliographical references and index.
1. Science–History. 2. Science–Cross-cultural studies. I. Title.
Q125.T724 2012
509–dc23
2011027400

ISBN: 978-0-415-78254-8 (hbk)
ISBN: 978-0-415-78255-5 (pbk)
ISBN: 978-0-203-14294-3 (ebk)

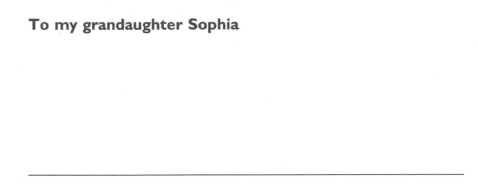

To my grandaughter Sophia

Contents

Chapter 1

What is science?

Imagine for a moment that you were an extraterrestrial approaching the planet Earth for the first time. What would you notice?

There are lots of candidates for answers to this question. You might, for example, exclaim over the presence of liquid water on the planet's surface—a rare phenomenon in the universe. You might wonder why the atmosphere was full of a corrosive, poisonous gas that the natives called oxygen. But my guess is that you would notice something else. Among the life forms present, you would notice that one species—the one that calls itself *Homo sapiens*—is somehow different. Alone among the millions of living things on the planet, this species has spread over the entire habitable surface, converted vast tracts of forest and grassland to farms, and built an interconnecting grid of massive cities. It has dammed rivers, built highways, and even come to dominate some of the natural chemical cycles that operate in the planet's ecosystems. While closely related to all the other life forms at the molecular level, this species is just ... well ... different.

Why?

I would suggest that the answer to this question lies in one simple fact. Human beings are the only life form on Earth that has developed a method that allows them to understand the universe around them (what we call science) and the ability to use that understanding to transform the environment to their advantage (what we call technology). It is these twin abilities, developed over millennia, that have allowed humanity to prosper as it has.

In fact, I will go so far as to argue that the really deep changes in the human condition—the ones that produce fundamental differences in our world—arise because of advances in science and technology. Let me give you two examples to back up this claim.

Forty thousand years ago our ancestors eked out a fragile existence as hunter-gatherers, harvesting the food that nature provided for them. Then, over a long period of trial and error culminating around 8000 BCE, some of them (probably mostly women) discovered that life didn't have to be lived that way. They observed the way that wild plants grew and realized that instead of being satisfied with what nature offered in the way of nourishment, they could plant seeds, tend the growing crops, and harvest the final product. The enterprise we call agriculture was born and the world has never been the same. The surplus of food allowed human beings to begin building cities, where arts and learning could grow. To be fair, it also allowed for the existence of standing armies, another, perhaps less welcome, aspect of modern life. But in any case, those early farmers, without writing, mostly without metal tools, used their observations of the world to change it forever.

Fast forward ten thousand years, to England in the latter half of the eighteenth century. This was a country poised to become the greatest empire the world had ever seen, a country with enormous social and class inequalities, and one whose major colony in North America was on the brink of declaring independence. Suppose you imagine yourself in London in 1776 and ask a simple question: what is going on in this country that will have the greatest impact on human life over the next couple of centuries?

I would suggest that if you wanted to answer this question you wouldn't go to the great universities or to the seats of government. Instead, you would go to a small factory near Birmingham, to the firm of Watt and Boulton, where the Scottish engineer James Watt was perfecting his design of the modern steam engine.

A word of background: there were steam engines in existence before Watt, but they were cumbersome, inefficient things. A two-story high engine, for example, developed less power than a modern chain saw. What Watt did was to take this cumbersome device and change it into a compact, useable machine.

Seen in retrospect, this was a monumental advance. For all of human history the main source of energy had been muscles—either animal or human—with small contributions from windmills and water wheels. Suddenly, the solar energy that came to Earth hundreds of millions of years ago became available, because it was trapped in the coal that was burned in Watt's steam engine. This engine powered the factories that drove the Industrial Revolution, the railroads that tied together continents, the cities where a greater and greater proportion of humanity spent their lives. The machines being built in that grubby factory were the agents of a fundamental change in the human condition. And whether you think this is a good thing (as I do) or a deplorable one (as has become fashionable in some circles), you can't deny that it happened.

Science and technology

In this book we will look at a number of other discoveries and developments that have had (or are having) the same sort of deep effect. The development of the electrical generator transformed the twentieth century, breaking forever the ancient link between the place where energy is generated and the place where it is used. The germ theory of disease changed the way medicine was done, producing unheard of lifespans in the developed world. The development of the science of quantum mechanics led to the digital computer and the information revolution that is transforming your life even as you read these words.

While it is indisputable that science and technology have changed our lives, we need to understand the differences between the two of them. In everyday speech they have come to be used almost interchangeably, but there are important distinctions that need to be made. As implied in the previous discussion, science is the quest for knowledge about the world we live in, technology the application of that knowledge to satisfying human needs. The boundaries between these two activities are fuzzy at best, with large areas of overlap—indeed, we will spend a good portion of Chapter 11 exploring in detail the process by which abstract knowledge is turned into useful devices. For the moment, however, we should just keep in mind the notion that these two terms refer to different sorts of processes.

Science and technology, then, go a long way toward explaining what the hypothetical extraterrestrial with which we started the discussion would observe. And this, of course, leads us to a number of interesting questions: what exactly is science, and how did it arise? How similar to modern science was the work of previous civilizations, and in what ways did their approaches differ from our own and from each other? Are there any activities that are common to every scientific endeavor? Before we get into a detailed description of the way that science is practiced in the twenty-first century, let's look at these historical questions.

The historical question

Later in this chapter we will describe the modern full blown scientific method in some detail, but for the moment we can picture it as a never ending cycle in which we observe the world, extract regularities from those observations, create a theory that explains those regularities, use the theory to make predictions, and then observe the world to see if those predictions are borne out. In simple diagrammatic form, we can picture the normal modern scientific method as a clockwise cycle (see overleaf).

One way of asking the historical question, then, is to ask what parts of this cycle various civilizations of the past used. The first two steps—observation of the world and the recognition of regularities—are pretty universal, and probably predate the appearance of *Homo sapiens* on the evolutionary scene. No hunting-gathering group would last very long if its members didn't know when fish would be running in a particular stream or nuts would be ripening in a particular forest. Indeed, we will argue in the next chapter that many pre-literate civilizations developed a rather sophisticated astronomy based on regular observations of the sky. The existence of structures like Stonehenge in England

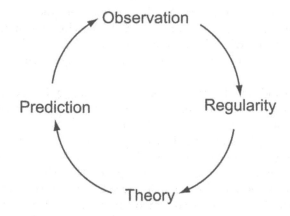

and the Medicine Wheels of western North America testify to this sort of development. An important lesson we learn from these sorts of structures is that it is possible to pass complex information about the natural world from generation to generation through the oral tradition, even in the absence of writing.

The absence of written records makes it difficult to know what, if any, theories these early peoples developed to explain what they saw. This situation changes when we look at the civilizations of Mesopotamia and Egypt. Here we run into a strange dichotomy. The Babylonians kept the best astronomical records in the ancient world—in fact, their data was still being used by Greek astronomers centuries after it was recorded. As far as we can tell, however, they seemed totally uninterested in producing a theory to explain their findings. It seems that if they were able to look at the data and figure out when the next eclipse would occur, they were satisfied. In terms of the cycle pictured above, they seemed to get off the train with regularities and not be interested in going any further.

The Egyptians are more typical. They told stories about what they saw in the sky, explaining the motion of the heavenly bodies in terms of the adventures of the gods. Whether this sort of explanation of nature constitutes a "theory" is a tricky question, depending as it does on how you define the word "theory." The point, however, is that once you explain any natural phenomenon in terms of the whims of the gods, you lose the power to make real predictions, since in principle those whims can change at any time. In this case, you are limited, as were the Babylonians, to relying on past regularities to forecast the future. As far as we can tell, this was the case for most of the advanced ancient societies we'll be studying.

The people who broke out of this mold were the Greek natural philosophers, who first began to construct theories based on purely naturalistic explanations of the world. By the first century CE, in fact, natural philosophers in Alexandria had put together a marvelously complex model of the solar system capable of making rudimentary predictions about things like eclipses, the rising and setting of the planets, and the time of the new moon.

During what are called the Middle Ages in Europe, the center of gravity for the development of science moved to the Islamic world (see Chapter 5) and progress was made in many areas—we will look specifically at mathematics, medicine, and astronomy. If you had to pick a date for the development of the modern scientific process, however, you would probably talk about the work of Isaac Newton in England in the seventeenth century (see Chapter 6). This is when the full blown scientific method outlined above made its appearance—the time when we went "all the way around the cycle."

Modern scientists tend to reserve the word "science" for the development that started with Newton (or, sometimes, with Galileo some decades earlier). In essence, they tend to regard what came before as a kind of "pre-science." Since this is common usage among my colleagues, I will use it, but in what follows I would urge you to keep in mind that one of the greatest failings of those who study history is to judge the past by the standards of the present. To my mind, the illiterate men and woman responsible for Stonehenge were every bit as good a set of "scientists" as my colleagues in any university science department of which I've been a member. The proper question to ask is not "How close did this ancient civilization come to what we do today?" but "What did they do and how did it fit in to their cultural life?"

Having said this, however, the modern scientific method can serve as a useful template that will help us organize the accomplishments of the various ancient civilizations we will study. It is useful, therefore, to examine this method in its current form, a subject to which we will devote the rest of this chapter.

The modern scientific method

Before we launch into this subject, I want to make a strong caveat—one that I will emphasize at the end of the chapter as well. Science is a human endeavor, carried out by human beings no different from the rest of us. One well known characteristic of human behavior is an aversion to blindly following rules. Like artists and musicians, scientists often delight in departing from the path of tradition and striking out on their own. Thus, what follows should be thought of as a list of elements found in most scientific work, more or less in the order in which they can normally be expected to appear. It should not be thought of as a kind of "cookbook" that all scientists follow at all times.

Observation

All science begins with observation of the world. It is important to point out that the idea that you can learn about the world by observing it, an obvious proposition to those of us living in secular, technology driven societies, has not been a universal given throughout most of human history. There are, in fact, many ways of approaching the problem of learning about the universe. In Chapter 4, for example, we will talk about the approach taken by many Greek philosophers, an approach in which the power of human reason, rather than observation, was the main tool for exploration.

We can see another way of approaching the world in the seemingly endless debate about the inclusion of creationism (or its latest incarnation, intelligent design) in the science curriculum in American public schools. On one side of this debate is the scientific community, relying on data gathered from the fossil record and modern measurements of DNA—data gathered, in other words, through observations of the world. On the other side we have people for whom a literal interpretation of the creation story in the Book of Genesis is taken as the inviolable, unquestionable, eternal word of God. For these people, the truth about the universe is contained in revered texts, and observations have nothing to do with it. For at least some creationists, it is impossible to imagine any experiment or observation that would convince them to change their minds. For people who think this way, in other words, you do *not* learn about the world by observation, but by consulting the sacred texts.

So with the caveat in mind that not all human societies would agree with the statement, we will begin our discussion of the scientific method with the following:

If you want to learn about the world, you go out and observe it

We will take this to be the first step in the development of modern science and, as we have argued, it was a step taken by many societies in the past. Having said this, however, we have to point out that there are many different kinds of "observation," each appropriate for a different area of science.

When most people think about what scientists do, they think about experiments. An experiment is a specific way of observing nature, usually under highly controlled (and somewhat artificial) circumstances. The basic strategy is to change one thing in a physical system and see how that system changes as a result.

A classic example of this approach to observation can be seen in the Cedar Creek Natural History Area near Minneapolis. There, scientists from the University of Minnesota have been studying the way that plant ecosystems respond to changes in their environment. They set up their experiment by having many plots of ground a few yards square. Every plot gets the same amount of rain and sunshine, of course, but the scientists can change the amount of other materials on the plots. For example, they can add different amounts of nitrogen to some plots and none to others, then watch the way the different plots evolve over the summer. This is a classic example of a controlled experiment. (For the record, the experiment I've just described showed that adding nitrogen increases the biomass in a plot, but lowers its biodiversity.)

In many sciences, this sort of finely controlled experiment can be done, but in others it cannot. An astronomer, for example, cannot build a series of stars to see the effect of adding a particular chemical element to the system, nor can a geologist go back and watch rock layers forming on the early Earth. Scientists in these sorts of fields have to depend more heavily on pure observation rather than experimentation. This doesn't affect the validity of the science, of course, but it's important to keep in mind that knowledge is acquired in a slightly different way.

Finally, as we shall point out in Chapter 12, the advent of the digital computer has introduced yet another meaning to the term "observation." Over the past couple of decades, as computers have gotten more powerful and our knowledge of the details of physical systems has grown, scientists have started to assemble massive computer programs to describe the behavior of complex systems—everything from the future of the climate to the evolution of ecosystems to the formation of planets. It is becoming more and more common for scientists to "observe" something like the formation of a planetary system by changing parameters in a computer program, in much the same way as the Minnesota scientists varied the amount of nitrogen in their plots. This sort of "observation" is usually referred to as "modeling" or "simulation."

Having made these distinctions, it is important to remember that whether scientists start their work with experiments, observations, or simulations, they always begin with a reference to what is seen in the external world.

Regularities

After you observe the world for a while, you come to an important realization: events do not happen at random. In fact, the world we inhabit is surprisingly regular and predictable. The sun always rises in the east and sets in the west, for example, and the days get longer and shorter with the seasons in a predictable way. So predictable is the world, in fact, that even ancient civilizations without writing were able to construct massive monuments like Stonehenge to mark the passage of time—a topic to which we'll return in the next chapter. Noticing and stating these regularities is the second important step in the scientific process. We have noted that most civilizations have reached this step in the scientific process. In fact, most of the activities we now characterize as "crafts" actually represent the accumulated experience of generations of people observing the natural world.

It is at the stage of finding regularities that those of us who teach science often begin to encounter a problem, because these regularities are often stated in a strange language—the language of mathematics—rather than in English. Mathematics is a somewhat artificial language that has the enormous advantage of having a high level of precision. Unfortunately, it is also a language that creates a high level of anxiety in many students. Let me say this about mathematics: there is nothing contained in any mathematical equation ever written by a scientist that cannot be stated in ordinary language (albeit not as elegantly). The translation of insights about the world's regularities into mathematics is no more mysterious than the translation of a poem from one language to another. Furthermore, all of the great truths of science can be stated without mathematics, especially since, as we shall see, most of them embody concepts with which we are already familiar from daily life. Consequently, in what follows, with very few exceptions, we will use ordinary language rather that mathematics.

Theories

After we have observed nature long enough to realize that it is regular and predictable, we can move on to what is perhaps the most interesting question in the scientific process. How must the universe be arranged so that it will produce the regularities we actually see? What does our experience, in other words, tell us about the nature of the world we live in? At this point, human thought leaves the realm of immediate experience and begins looking for a deeper meaning to what we see. I will call this process the construction of a theory.

A word of warning: there has been (and continues to be) a great deal of debate among philosophers of science about the precise definition of the word "theory," with different camps placing different constraints on how the word should be used. In addition, there is an unfortunate confusion in the minds of many people because in colloquial language, the word "theory" is sometimes taken to be synonymous with "unsupported guess." You see this occasionally, for example, in the debates about including Creationism on the public school curriculum. Evolution, in these debates, is often derided as "just a theory"—a statement that emphasizes the difference between the way scientists and the general public use the word.

Because of these problems, let me make a brief aside at this point to make it clear what I (and most scientists) mean when we talk about a theory. I have to start this discussion with a somewhat unusual observation: as good as scientists are at learning about how the universe works, they are really pretty bad at naming the things they discover. In the next section, for example, we will see that the cornerstone—the absolute bedrock—of science is a relentless testing of ideas against observation. It would be very nice, then, if an idea being proposed for the first time were called an "hypothesis," and after it had been verified a thousand times it became a "theory," and, finally, after a million verifications it became a "law." Unfortunately, it just doesn't work that way. Whatever name an idea gets when it is first introduced stays with it, no matter how well it is supported by observation subsequently.

Let me give you an example: In Chapter 6 we will talk about the work of Isaac Newton and, in particular, his development of what was called the "Law" of Universal Gravitation. For several centuries this was our best explanation of the phenomenon of gravitation, and it is still used today when we want to send a space probe to a distant planet. In the early twentieth century, Albert Einstein developed a deeper explanation of gravity, which goes by the name of the "Theory" of General Relativity. This "Theory" contains within it, as a special case, Newton's "Law" of Universal Gravitation. Thus in this example the thing we call a "law" is actually less general and less well validated than the thing we call a "theory." This is a dramatic example of the kind of deficiency in the scientific naming process I mentioned above, but far from the only one.

Because physics was the first of the sciences to develop in its modern form, there has been a tendency for subsequent scholars to try to cast their science in the mold first established by physicists. There is even a tongue-in-cheek term— "physics envy"—that academics use to describe this phenomenon. But just as there are different types of "observation" appropriate to different branches of science, so too are there different types of theories.

Physics tends to be an exact science, driven by precision experiments and highly quantitative theories. In the best cases, in fact, the results of theoretical calculations and laboratory experiments can agree to more than ten decimal places! The theories of physics, as we shall see in Chapter 7, tend to be stated in rigorous mathematical terms (although, as stressed above, they *can* be stated in words as well). This gives them an aura of precision that can sometimes be deceiving, but which characterizes one type of ideal for a scientific theory.

At the other end of the spectrum are theories that are less quantitative, theories that describe general trends and processes. One example of this type of theory is Darwin's original statement of the laws of natural selection, which we will discuss in Chapter 8. Rather than making precise predictions ("42.73 percent of this type of animal will survive long enough to have offspring"), this theory enunciates general rules about how populations develop over time ("In this situation, animal A is more likely to have offspring than animal B"). Many theories in the historical sciences, such as the theory of plate tectonics (our modern view of the Earth's structure) are of this type. (Having said this, I have to point out that since Darwin, many highly quantitative developments have been added to the theory of evolution. It is not at all unusual, for example, to hear a modern paleontologist using DNA analysis or complex mathematical calculations to buttress a thesis about the development of a particular organism.)

This difference between the different kinds of theories in the sciences actually fuels one of the more interesting debates among scholars today. The debate centers on the question of whether the universe is deterministic or contingent. As we shall see in Chapter 6, the work of Isaac Newton led to a particularly mechanical view of the universe, one in which observed phenomena were thought of as being analogous to the hands of a clock. In this picture, the gears of the clock are the laws of nature, and if we only knew enough about the gears we could predict exactly where the hands would be at any time in the future. The universe, in this picture, is completely deterministic. On the other side, people like the paleontologist Steven Jay Gould have argued that the universe (at least in its biological aspects) is contingent—"Run the tape again" he argued, "and you get a completely different tune."

This is one of those deep questions that it's easy to ask and very difficult to answer. For the record, my own conclusion that that the universe is a lot less contingent than Gould would have it, and probably a little less deterministic than Newton thought.

Prediction and verification

Once we have an idea about how the universe works, we are ready to take the final step in the scientific process—a step, I argue, that separates science from all other human intellectual activities. The basic idea is that we look at our theory and ask a simple question: does this theory predict anything that I haven't seen yet? If the theory is worth anything, it will suggest many such observations of as yet unobserved phenomena, and this, in turn, leads to new experiments and observations. In fact, it is this relentless testing of theories against actual observations of the natural world that is the distinguishing mark of the scientific enterprise. To put it bluntly,

In science there are right answers, and we know how to get them.

There are several points that can be made about this statement. We started the scientific process by observing nature and, in the end, we finish it in the same way. Science begins and ends by comparing our ideas against the realities of the natural world, by observing natural phenomena. It is the presence of this impartial outside arbiter of ideas that makes science different from other fields. Every scientist (the author included) has had the experience of starting an argument with reasonable hypotheses, following impeccable logic to an unquestionable conclusion, only to have experiment show the whole thing to be wrong. In the end, it doesn't matter how well you frame your arguments or how much status you have in the scientific community—if the data doesn't back you up, you are wrong. Period.

Once we realize how important this verification process is, however, we have to recognize that it can have two possible outcomes. It may be that our predictions are borne out by observation. Oddly enough, this is not the outcome most scientists hope for when they begin an experiment, because such a result really doesn't add much to our store of knowledge. Furthermore, in the scientific community there is normally much more of a cachet attached to disproving a theory than affirming it. Nevertheless, a positive outcome means that our theory has been confirmed. In this case, scientists will usually look for another prediction that they can test.

The other alternative to the verification process is that the prediction may not be borne out. What happens in this case depends on the status of the theory being tested. If it is a new theory, being tested for the first time, scientists may simply conclude that they have gone down a blind alley, abandon the theory, and try to construct a new one. If, on the other hand, the theory has enjoyed some success in the past, scientists will look for ways to modify and extend it to accommodate the new finding. Instead of asking "What other theory can we build?", in other words, they will ask "How can we modify this theory to make it more complete?" We will see many examples of both of these processes as we examine the historical development of science.

Just so that we have a concrete example of the prediction-and-verification process in mind, let's turn our attention to a series of events that followed Newton's development of his mechanistic view of the universe. In one fell swoop, he had reduced millennia of astronomical observations to simple consequences of a few deep physical laws. In his orderly, clockwork universe there was only one flaw—the occasional appearance of comets.

Think for a moment about how a comet must have appeared to people like Newton. The orderly progression of the planets through the sky—motion we compared to the movement of clock hands above—is suddenly interrupted by the appearance of a strange light in the sky. The comet hangs around for a while, then disappears. What was that all about?

Edmond Halley was a distinguished scientist who settled down as Astronomer Royal after an adventurous youth. A friend of Newton, he decided to tackle the comet problem. According to one story, he was having dinner with Newton and asked his friend a simple question: if comets were material bodies affected by gravity like everything else, what shape would their orbits be? Newton had thought about this problem, and told his friend that comets would have to move in an ellipse. Armed with this knowledge, Halley examined data on 26 historical comets to determine their orbits, and discovered that three of those comets moved in exactly the same elliptical path.

Flash of insight. That wasn't three separate comets in the same orbit—it was one comet coming back three times. Using Newton's Laws, Halley calculated when the comet would be seen again and made his prediction—it would return in 1758. Sure enough, on Christmas Eve, 1758, an amateur astronomer in Germany turned his telescope to the sky and saw the comet. This event, which historians call the "recovery" of Halley's comet, can be taken to be symbolic of the development of the modern scientific method.

I would like to make a couple of points before we leave Halley. In the first place, we can imagine a scenario in which the comet failed to appear—things didn't turn out that way, but they could have. This means that Halley's prediction could have been wrong. In the language of philosophers, the

Newtonian theory is "falsifiable." (In the American legal system, this same property of scientific ideas is called "testability".) It is important to realize that a falsifiable statement can be either true or false. For example, the statement "The Earth is flat" is a perfectly falsifiable scientific claim that happens to be false. The statement "The Earth is round" is also falsifiable, but happens to be true. All real scientific theories must be falsifiable (which is not the same as saying they must be false). If you have a theory that cannot possibly be proved wrong, as is the case in some versions of Creationism, it simply is not science.

I can't leave Halley, though, without quoting a statement he made when he predicted the return of his comet:

> Wherefore, if (the comet should return again about 1758, candid posterity cannot refuse to acknowledge that it was first discovered by an Englishman.

The growth of science

Most of the time, most scientists are engaged in pursuing their craft in more or less the order given above, beginning and ending their work with observations of nature. As I intimated above, however, these steps are not a cookbook, and there are times when scientists joyfully "break the rules." Perhaps the most famous example is Albert Einstein's development of the Theory of Relativity, which began not with observation, but with a deep analysis of a fundamental contradiction between two different areas of physics. Once enunciated, though, the theory had to go through the same prediction-and-verification process as everything else. (We'll talk about relativity in more detail in Chapter 9).

Just as there are different types of "observation" appropriate to different areas of science, there are different ways that areas of science advance. One way that science can change is by the simple replacement of one theory by another. When Nicolas Copernicus first put forward the idea that the Earth orbited the sun rather than standing stationary at the center of the universe, his ideas eventually replaced the prevailing theories about geocentrism. Generally speaking, this kind of replacement process tends to occur early on in the development of a science, while there are still a lot of unknowns and a lot of room for theorists to maneuver. The last time this sort of wholesale replacement happened was in the Earth sciences in the 1960s, when the modern theory of plate tectonics, with its mobile continents, replaced the old theories of the fixed Earth.

Once a field of science reaches a certain level of maturity, however, a different type of change starts to predominate. Instead of replacing old theories with new ones, scientists extend existing theories, adding new material without abandoning the old. As we shall see in Chapter 9, the great advances in physics in the twentieth century (relativity and quantum mechanics) do not replace Newtonian physics, but extend it to new areas, areas where it was not originally known to apply. We shall see that if we apply the rules of quantum mechanics, derived for the world of the atom, to large scale objects, those rules become identical to Newton's Laws. Thus this type of change in the sciences can be thought of as being analogous to the growth of a tree. New material is always being added on the periphery, but the heartwood remains unchanged.

Another way of thinking about this picture of incremental growth in the sciences is to go back to our basic premise that science begins and ends with observation. Every great law of science is based on observation, and is only as good as the observations that back it up. Newton's picture of the universe is massively supported by observations of normal sized objects moving at normal speeds, but until the twentieth century we had no observations of objects on the scale of atoms. When those observations came in, they led to a new field of science—quantum mechanics—but did not contradict the old constellation of observations. Thus, Newton's Laws remain the heartwood, and we still use them to build bridges and airplanes, even though we understand that they can't be applied outside of their original area of validity.

And this brings us to another important aspect of the scientific process. The pragmatic methods I've described can be thought of as a way of finding progressively more exact approximations to the truth, but they will never get us to Truth. Every law of science, no matter how venerable, can, in principle, be proved wrong by a new observation. Such a turn of events is surely very unlikely, but the requirement of falsifiability demands that it be possible. It is the nature of science that all truths are tentative, subject to the results of future observations.

Finally, I would like to end this introduction to the scientific process by discussing an aspect which has been the subject of academic debate in recent years. This debate has to do with the role of social norms in the development of scientific theories. On the one side we have working scientists who believe that they are finding better and better approximations to reality through their work. On the other side are philosophers and sociologists of science, who argue that scientific theories are, in fact, the result of what they call social construction. In its most extreme form, this argument becomes a kind of solipsistic exercise—in essence, the argument that an observed regularity in nature has no more intrinsic meaning than the convention that a red light means "stop" and a green light means "go."

It shouldn't surprise you that working scientists, with their emphasis on observation and verification, who heard about these arguments disagreed violently. This led to an episode called the "Science Wars," which was basically a debate between the views outlined above. (I should say, however, that most scientists never heard of this debate, which made many observers wonder if you can really have a "war" when one side isn't aware that it's going on.)

The basic issue in the "Science Wars" was the extent to which social structures affect the results that scientists derive. That society affects science (and that science affects society) can scarcely be denied. The real issue is the extent to which social effects can determine the results of the scientific process outlined above.

There can be little dispute that in the short term, social influences can have a large effect on scientific development. Governments, for example, can encourage certain areas of research by funding them, and discourage others— even make them illegal (as has been done for human cloning in many countries, for example). In rare instances, governments can even try to suppress scientific results. (If you want to see a particularly egregious example of this, Google "Trofim Lysenko" to learn how Josef Stalin delayed the development of modern biotechnology in the old Soviet Union by several generations.) In addition, there are fads and trends in the sciences themselves that can influence the way research is done and the way it is interpreted.

In the long run, however, these kinds of effects are ephemeral. As we emphasized above, in the end what matters in science is verification through observation. No amount of government intervention or social pressure could have removed Halley's comet from the sky on that day in 1758, for example. Even cases of scientific mistakes (which happen) and scientific fraud (which also happens, though less frequently) are eventually uncovered by the scientific process.

There is only one place where social influences can have an important effect in the scientific process, and that is in the construction of theories. Scientists, after all, are members of their societies. At any given time, in any given society, there are some ideas that simply can't be thought, not because they are forbidden, but because they are just outside of the mental landscape of the time. For example, Isaac Newton could no more have conceived of the theory of relativity than he could have written rap music. In this sense, and in this sense only, we can speak of science as being "socially constructed."

In any case, the process outlined above represents the way that science works in its mature form. In what follows we will see how the elements of the scientific process developed in cultures around the world, reaching its modern form in western Europe in the 1600s. From there, we will trace its spread, first to places like Russia and America on the periphery of Europe, and then to the entire globe.

Chapter 2

Astronomy: the first science

At first it may seem a little strange that the first science to develop was astronomy. After all, the stars and planets are far away, and have little influence on the daily life of the average person. Why would a group of hunter-gatherers or primitive farmers care about the nightly display in the sky?

This attitude, widespread and understandable in the twenty-first century, ignores an important historical fact. The technology that led to widespread street lighting wasn't developed until the end of the nineteenth century, and widespread electrification didn't follow in most places for another fifty years. This means that when it got dark throughout most of human history, people were presented with the full glory of the sky, unalloyed by the presence of street lights.

Think about the last time you were out in the country at night, far from city lights. Do you remember how close the stars seemed, how immediate? For most of human history, this is what people lived with day in and day out. *Of course* they were aware of what was going on in the sky—it was laid out for them every night. (If you've never had the experience of seeing the dark night sky yourself, I hope you have it soon. It's not to be missed.)

Living in cities, surrounded by artificial lights, the sky has receded from our consciousness. On a clear night, you might see the moon and perhaps a few stars, but that's about it. There's nothing in the sky to compare with an ordinary street light, much less a blazing location like Times Square. To us, the sky is, at best, an incidental part of nature.

When I want to explain the difference between the way the sky was seen throughout history and the way it is seen now, I often refer to a typical urban experience: Rush Hour. I daresay that these words conjured up instant images in the minds of most readers—probably an image of long lines of cars inching along on a highway. The point is that no one ever sat you down and explained what rush hour was, and you never took a course titled "Rush Hour 101." Rush hour is just a part of your life if you live in a city, and you know that there are certain times of day when you do not drive in certain areas if you can help it. Rush Hour simply seeps into your consciousness if you live in an urban environment. In the same way, I suggest, the movement of bodies in the heavens seeped into the consciousness of our ancestors.

In addition, there is one very practical bit of knowledge that can be gleaned from observing the motions of objects in the sky: they can provide us with a calendar to mark the progression of the seasons. Again, the concept of living without a calendar seems very strange to the modern mind. Since 1582, we have had the Gregorian calendar, which does a pretty good job of keeping the seasons in line with the calendar dates, and, if need be, the scientists at the Naval Research Laboratory in Washington DC can insert a leap second or two from time to time to keep things in tune.

To understand why the construction of our current calendar was so long in coming, you have to realize that there are two "clocks" provided by nature to help us keep track of the time. One of these is the rotation of the Earth, which provides a reliably repeating event (such as sunrise) every 24 hours. The other clock involves the revolution of the Earth in its orbit around the sun, a period we call a year. In the modern view of things, it is the position of the Earth in its orbit that determines the seasons. On the other hand, counting sunrises or some similar event is the easiest way to keep track of the passage of time. The role of a calendar is to reconcile these two clocks—to use the easy counting of days to gauge the position of the Earth in its orbit. (For completeness, I should point out that there is a third "clock"—the recurring phases of the moon. This "clock" gives rise to lunar calendars, which we will not discuss here.)

The importance of a calendar for an agricultural society is obvious. It is important to know when crops should be planted, and the daily weather isn't always a good indicator. When I want to make this point, I talk about the birth of my oldest daughter. It was February in Virginia, but an unusually warm day—I remember walking around in shorts, buying a collection of all the papers published on that day for souvenirs. On my daughter's third birthday, however, I spent the day digging visiting friends out of huge snowbanks. The Earth was at the same point in its orbit both times, but the weather was very different. Had I been fooled by the warmth and planted my garden that first February, the plants would have been killed off by cold later in the month. This wouldn't have been a disaster for me—I could always go to the supermarket—but had I been directing an early agricultural settlement, it could have meant starvation for the whole community.

The role of a calendar is to use recurring events in the sky to determine (again, in modern language) the position of the Earth in its orbit around the sun and, thus, to keep track of the seasons. The most obvious marker events have to do with the position of the sun in the sky at different times of the year. Even confirmed city people are aware that the days are longer in summer than in winter, and may also be aware that the sun is lower in the sky during the cold months. Our modern understanding is that this phenomenon results from the tilt of the Earth's axis of rotation, but ancient observers could hardly have been aware of this explanation.

Imagine, if you will, what the year looked like to them. As winter deepened, the sun would rise later and later each day. Furthermore, each day the sun would come up a little farther south than the day before. And then, one bitterly cold morning, the miracle happened! The sun reversed itself and rose a little to the north and the days started to get longer. Small wonder that this event, the winter solstice, was celebrated by lighting fires throughout the northern hemisphere. (Didn't you ever wonder why we put lights on Christmas trees?)

It's not hard to imagine an oral tradition starting, even among people without writing, to keep track of this event. You can easily imagine a gray bearded elder telling his grandchildren that if they stood on a certain spot, they would see the sun come up over a particular notch in the hills on that magical day. This sort of event, repeated in many places around the world, is an example of the discovery of regularities we discussed in the last chapter.

It is also a way to locate the position of the Earth in its orbit around the sun. We have to remember, though, that the way we look at the sky today, as a collection of objects obeying the cold, impersonal laws of physics, is not the way the sky would have been seen by our ancestors. Take a simple sequence of events like sunrise and sunset. To us, this is a simple consequence of the rotation of our planet. If we want to picture it, we can imagine a rotating globe with a light shining on it from one direction. Places on the globe move into the light as the sphere rotates, and move back into darkness later. Simple enough. But the movement of the sun would certainly not have looked that way to the first astronomers.

In many ancient civilizations, in fact, the sun (or possibly the Sun God) was thought to die at sunset, perhaps to wander in the underworld for a time, and then be reborn at sunrise. In Egyptian mythology, for example, the Sky Goddess Nut gave birth to the sun each morning. The fact that the sun went down and died in the west is said to have given rise to the fact that all of the great burials of Egypt, from the Pyramids to the Valley of the Kings, were located on the west bank of the Nile. In fact, in ancient Egyptian, the phrase "went to the west" played roughly the same role as the phrase "passed away" does in our own culture. My own favorite sun story was part of Aztec mythology. In this account, the Sun God was born each day and rose to the zenith, only to be dragged down and killed at sunset by the spirits of women who had died in childbirth. Scary!

I mention these differences between the modern and ancient views because they illustrate some of the dangers of looking at ancient science through the lens of modernity. As a physicist, it is easy for me to look at ancient astronomy as nothing more than precursor of our own impersonal view of the universe. Fair enough, but that isn't what the ancient astronomers thought they were doing. Like me, they were trying to deduce important things about the universe by observing the sky. The fact that they saw the actions of the gods where I see the operation of Newton's Law of Universal Gravitation isn't what's important—what matters is that we are both observing the universe and trying to understand what it means.

It is only fairly recently that we have started to acquire a real appreciation of the level of sophistication of ancient astronomers. Since the 1960s, however, a thriving field of research called "archeoastronomy" has developed. The goal of archeoastronomy is to understand both the astronomical knowledge of ancient civilizations and the role that that knowledge played in ancient cultures. Practitioners of the field tend to fall into one of two camps—the astronomers, who tend to concentrate on the science, and the anthropologists, who tend to concentrate on the culture. In the beginning, this dichotomy had some amusing consequences. For example, at one of the first formal academic meeting devoted to the new discipline, held at Oxford University in 1981, the two camps thought themselves to be so different that the conference proceeding were actually published in two separate volumes. The two different approaches to archeoastronomy were then named for the colors of those published books, with "Green Archeoastronomy" basically denoted the astronomical, and "Brown Archeoastronomy" the cultural and anthropological. Needless to say, the two camps have made good progress since then, with each side learning a little about the other's discipline.

The birth of astronomy

We will never know about the first person who began a serious observation of the night sky, since he or she undoubtedly belonged to a culture without writing. The advent of archeoastronomy, however, has shown us that we can glean some information about the beginnings of astronomy by examining the buildings early astronomers left behind. As to the motives of these early people, the stones themselves are mute. The most common explanation, given above, is that at least one function of astronomy in the early days was the establishment of a calendar. This couldn't have been the only function, however, because, as we shall see below, people who did not practice agriculture developed astronomy as well as those who did.

Even though we can't assign a firm date to the birth of astronomy, we can pick a well known structure to symbolize the event. I would like to nominate Stonehenge, in southern England, to play this role. This great circle of standing stones in southern England is one of the best known ancient monuments – so much so, in fact, that its image is used as one of the standard computer screen wallpapers in Windows. Construction on Stonehenge began in about 3100 BCE, and the choice of this monument for our discussion has the added advantage of illuminating the birth of archeoastronomy. This field can be said to have started with the publication, in 1965, of an analysis by British born astronomer Gerald Hawkins of the alignments of stones in the monument.

Before going into Hawkins' work, let me take a moment to describe the monument itself. The main part consists of two concentric stone circles, with the outer circle having a post and lintel construction (i.e. it has two upright stones with a third laid across the top). Some of the stones are quite large—they can weigh up to 50 tons, so moving them was not a trivial project. Roughly speaking, the monument was built in stages and finished around 2000 BCE.

There are all sorts of legends about Stonehenge—that it was built by the Druids, that it was built by Julius Caesar, that the stones were flown from Ireland by Merlin the Magician (or even by the Devil himself). In fact, the stones had been in place for two thousand years before the Druids ever showed up. As best we can tell, the monument was built by successive waves of Neolithic people—people who had neither writing nor metal tools.

Hawkins grew up near Stonehenge, and used to wander around the stones as a boy. He realized something important about the way the stones were arranged. When you stand at the center of the monument, there are certain very well defined lines of sight created by the stones. It's almost as if the builders are telling you "Look this way—here's where something important is going to happen."

Hawkins wondered where those lines of sight pointed. As a professional astronomer, he had access to newly available digital computers at MIT and was able to show that many of those lines pointed to significant astronomical events. The most famous of these marks the summer equinox—the beginning of summer. On that day a person standing in the center of the stone circle will see the sun come up over a pointed stone (called the heelstone) set in the field about 100 feet away. In fact, modern day Druids don their white robes and travel to Stonehenge to witness this event every year.

Hawkins' computers found other significant alignments, including the mid-summer sunset, the midwinter sunrise and sunset, and the rising of the full moon nearest these events. He also argued that it would be possible to use various parts of the monument to predict eclipses, but this assertion has not been widely accepted by scholars.

In terms of our modern understanding, the rising of the sun over the heel-stone at Stonehenge marks the time when the Earth is in a specific place in its orbit each year. It marks day one of the new year, and it is a simple matter to count the days from that point on. Thus, the notion of Stonehenge as a "calendar" is one you see often these days.

A word of explanation: you need the annual re-setting of the day counting because the number of days in a year isn't an integer. We're used to thinking of the year as having 365 days, but in fact it has 365.2425 days. This means that if you just counted 365 days from the heelstone sunrise, the Earth would actually be about a quarter of a day (6 hours) back in its orbit from where it was when you started counting. Keep up the counting for four years and it will be a whole day back—this is why we have leap years. By starting the counting for each year from the heelstone sunrise, the proprietors of Stonehenge would have kept their calendar in line with the seasons.

It is interesting that when these ideas about Stonehenge began to gain currency, there was a reluctance in some quarters to accept the notion that people with such a primitive technology could actually have carried out such a project. In a famous exchange, a scholar at an academic meeting excoriated his colleagues for suggesting the "howling barbarians" would be capable of such a thing. As time went on and, more importantly, as it became recognized that archeoastronomical sites could be found all around the world, this sort of opposition faded away. As I hope I made clear above, it is not at all surprising that people without writing should have acquired the astronomical know-how needed to build this sort of structure. The really remarkable thing is that they had the social organization capable of completing a project of this magnitude.

Before leaving Stonehenge, we have to make one point. Just because Stonehenge functioned as an astronomical observatory, that doesn't mean that it was *only* an astronomical observatory. The midsummer sunrise, the day when the sun is at its highest in the sky, was undoubtedly a moment of religious significance—an early version of the "sunrise services" you can still see celebrated at Easter. Furthermore, the area around Stonehenge contains one of the richest collections of Neolithic monuments in the world. Scholars have suggested that Stonehenge may have served as a site for significant burials, as part of an interconnected network of holy places, or as a place of healing, like modern day Lourdes. This work remains speculative because without written records it's really hard to tease out this kind of information.

But whatever other functions it may have had, there can be little doubt that (in modern terms) Stonehenge also played the role of astronomical observatory, keeping the seasons in line with the counting of the days. It has, as mentioned above, become something of a cultural icon for us. If you are ever in the town of Alliance, in western Nebraska, for example, you will be able to visit Carhenge, a reproduction of Stonehenge built from used 1972 Cadillacs—another experience not to be missed!

Alignment analysis

The kind of analysis that Hawkins did on Stonehenge is an example of what can be termed "alignment analysis." The idea is that you take an ancient structure and draw lines through its significant features—between two large upright stones, for example. You then see if any of those lines point toward an important astronomical event, such as the midsummer sunrise. If you get enough significant alignments, you are probably safe in assuming that the monument was built with astronomy in mind.

Once this technique was shown to work at Stonehenge, scholars quickly realized that many ancient structures in both the Old and New Worlds were aligned with events in the heavens. This insight was particularly important in analyzing things like the stone circles that dot northern Europe. Since we have no written records that tell us how these structures were used, we are more or less forced to rely on the stones themselves to tell us their story. The situation in Central and South America is a little different, since we have the written records of the Spanish Conquistadors to give us some sense of how the indigenous religions functioned.

Let me discuss one example of a building constructed to align with a specific celestial event. In Nubia, in southern Egypt, there is a great temple at a place called Abu Simbel. You have probably seen pictures of it—it's the one that has four huge statues of a seated pharaoh in front. This temple was built by Ramses II, arguably the greatest of the Egyptian pharaohs, probably (at least in part) to serve as a warning to potential invaders trying to enter Egypt from the south. Like all Egyptian temples, it consists of a series of chambers laid out in a line, with the chamber farthest from the entrance serving as a "holy of holies." Most temples are built from stone on flat ground, but Abu Simbel was actually carved into solid rock in a cliff overlooking the Nile.

The direction of the axis of the temple was carefully chosen, however, so that on two days a year, October 20 and February 20, the rising sun shines through the entire 180-foot structure and illuminates the statues in the innermost chamber. Since these days are the pharaoh's birthday and coronation day, it's pretty clear why this alignment was chosen. In this case, the existence of written records makes clear what the significance of the temple's alignment is.

As an aside, I should mention that when the Aswan High Dam was built in the 1960s, the rising waters in what was eventually called Lake Nasser threatened to submerge this magnificent temple. In an amazing effort, international teams of engineers disassembled the entire structure and put it together again on higher ground, even building an artificial cliff to replace the original.

Although the argument from alignment can give us powerful insights into the astronomical knowledge of ancient people, it is an argument that needs to be used with some caution. For one thing, if a structure is complex enough (as Stonehenge certainly is), some of the possible lines you can draw will point to an astronomical event, just by chance. Imagine, for example, drawing lines connecting all the entrances to your favorite athletic stadium. There are probably hundreds of such lines, and it would be amazing if at least one of them didn't point to something important. This means that even if you find a few alignments in a structure, you will probably need an additional argument if you want to convince skeptics that the structure had an astronomical use.

When I want to make this point, I tell people about the astronomy department at the University of Virginia in the 1980s and 1990s. During this period, the department was housed in a building that had originally been the Law School (and has since become the Department of Environmental Sciences). There was a long corridor on the third floor of this building, where most of my friends' offices were located. As it happened, at sunset on April 13, the sun was in a position to shine through a window at the end of the building and penetrate down the entire corridor. Since April 13 is the birthday of Thomas Jefferson, the founder of the University of Virginia, it's not hard to see that some archeoastronomer in the future might attribute an element of design to this alignment when, in fact, it is simply a coincidence—one of those things that happens now and then.

In the case of Stonehenge, the alignment of the heelstone is clearly significant—the placement of that stone away from the main circle gives it a special significance. It's clear that the fact that the heelstone was put where it is constitutes an example of the phenomenon I referred to above, where the builders are telling us "Look this way."

Monuments in the New World

Let me close this discussion of archeoastronomy by talking about two very different kinds of structures found in the New World, products of very different types of culture. One is the "Observatory," or El Caracol, at the Mayan center at Chichen Itza, in the Yucatan Peninsula in Mexico. The other is the Medicine Wheel in the Big Horn Mountains of Wyoming.

Construction on the Observatory probably began in the ninth century, at the end of the Mayan classical period. It's a two-story building, with a square base and the upper story being a rounded dome reminiscent of a modern observatory. The Spanish name, El Caracol ("The Snail"), comes from the winding stone staircase on the inside. The alignments of windows in the upper story point toward sunset on the summer solstice, the winter solstice sunrise, and to the northernmost and southernmost positions of the planet Venus. (Venus played a special role in Mayan divination, and its period was the basis for one of the calendars they used.) Part of the wall of the upper story has collapsed, but scholars suggest that it contained windows that pointed, for example, to events like the midsummer sunrise. Astronomer Anthony Aveni of Colgate University has calculated that 20 of the 29 alignments in El Caracol point toward significant astronomical events—a ratio that is well above chance.

The Observatory sits in the middle of the complex of Chichen Itza, probably best known for its massive stone pyramid. It is clearly the product of a technologically advanced agricultural society. The medicine wheels of North America are very different. They are found throughout the mountainous west in the United States and Canada, with the largest and best known being in the Big Horn mountains of northeastern Wyoming. Unlike any of the monuments we've discussed to this point, this medicine wheel is a rather primitive structure, consisting of little more than stones laid out of the ground. The basic structure is a wheel with spokes. There is a small stone circle, or cairn, at the "hub" of the wheel as well as six other cairns located around the periphery. The cairns are big enough to accommodate a seated observer. The whole thing is about 80 feet across, and the stones themselves are not very large—most could easily be carried by one man.

The Wyoming wheel is located in a spectacular setting, on a flat plateau just at the edge of the mountains. This location affords an unobstructed view of the sky, as well as the Big Horn Basin below. Astronomers have noted several significant alignments that could be seen by people sitting in the various cairns. Two are for the summer solstice sunrise and sunset, respectively. (There are no winter alignments, since the wheel would be under several feet of snow at that time.) In addition, there are alignments that point toward what are called heliacal rising of significant stars (Aldeberan, Rigel, and Sirius). A "heliacal rising" corresponds to the first time a star is visible above the sun before dawn, and would be another indicator of the seasons.

The interesting thing about medicine wheels—there are almost 150 of them—is that the people who built them were nomadic hunters, not agriculturalists. The fact that these sorts of people could not only acquire sophisticated astronomical knowledge but incorporate that knowledge into structures, no matter how primitive, is probably the best evidence I can present about the ability of early people to study the sky.

Naked eye astronomy: Babylon and China

The telescope wasn't invented until the mid 1600s, so by default all earlier celestial observations had to be made with the unaided human eye. Many ancient civilizations kept written records of astronomical observations, and these records give us a pretty clear idea of what they considered important. We will close these chapters by considering astronomy in two widely separated cultures: Babylon and China.

Civilization developed between the Tigris and Euphrates rivers, in what is now Iraq, well before 3000 BCE, and naked eye astronomy was carried out in this region for several thousand years. We're fortunate, because not only did these civilizations have writing, but they recorded their observations on a particularly durable medium—clay tablets. (We'll encounter this again in Chapter 3, when we discuss the Babylonian numerical system.) Basically, numbers and words were recorded by pressing a wedge-shaped stylus into soft clay—a method of writing referred to as "cuneiform," from the Latin "cuneus," or wedge. Hardened clay tablets can survive fire, flood, and the collapse of buildings, so our knowledge of Babylonian astronomy is pretty extensive.

Looking at Babylonian astronomical records can be a strange experience for a modern scientist. On the one hand, there are centuries worth of data on things like the time of the new moon, eclipses, and the positions of stars and planets in the sky. This data would turn out to be very important for later Greek astronomers like Hipparchos and Ptolemy (see Chapter 3). These sorts of activities were doubtless used to keep the calendar in line with the sky, as discussed above. More importantly, though, the Babylonians had a deep seated belief that events in the heavens could influence events on Earth. Consequently, much of their astronomical writing is concerned with finding omens, and with statements of the form "If an event X occurs in the sky, then there is a danger that Y could happen on Earth."

A word of explanation: omens are not predictions. An omen does not say that if X is seen, then Y is sure to happen. It's really a statement that if you see X, then you'd better be careful about Y. Think of it as being analogous to a modern physician telling a patient that if he or she doesn't do something about cholesterol levels, there is a danger of heart disease. Not everyone who has high cholesterol has a heart attack, but the cholesterol level puts you in a higher risk category. A Babylonian omen based on events in the sky would be the same sort of thing, so kings would often consult the stars on questions like whether or not to start a military campaign at a particular time. Again, you could still lose if the signs were good, but the odds were more in your favor than they would be if the signs were bad.

China is another civilization with a long standing record of naked eye astronomy. We have extensive records going back to the Warring States period in the fourth century BCE, and fragmentary records going back much farther. In China, the activities of astronomers seem to have been closely tied to the functioning of the government—astronomers would, for example, be expected to produce a new calendar for each new dynasty of rulers. They were also expected to predict important events like eclipses. Indeed, there are records from around 2300 BCE that tell of two astronomers being beheaded because they failed to predict the right time for an eclipse.

There are several aspects of Chinese astronomy that catch the attention of modern scientists. They kept meticulous records of events that we call super-novae (the Chinese called them "guest stars"). These events are connected with the explosive death of large stars. To an observer on Earth, they manifest themselves by having a star appear suddenly where there was no star before, hang around for a few months, and then disappear. Modern astronomers can detect the sites of ancient supernovae by looking for the expanding clouds of debris they generate, which means that the Chinese records can be corroborated by modern techniques. Perhaps the most famous of these is the so-called Crab nebula in the constellation Taurus, which appeared in 1054 CE. This was observed by not only the Chinese, but by Japanese and Arabian astronomers, and perhaps by Native American in Chaco Canyon, New Mexico, as well. (Oddly enough, we find no mention of it in European records.)

One aspect of Chinese astronomy that played a role in modern science concerns the appearance of sunspots. Under normal circumstances, humans cannot look at the sun with the unaided eye (you should *never* try to do this). In northern China, however, there are dust storms which block enough light to allow observation of the solar disk, and from about 200 BCE onward we find records of spots seen in the sun. The statements are in the form that "a three legged crow (or a star or a tool used for measuring rice) was seen within the sun," which is interpreted as the appearance of a sunspot (or, more likely, a sunspot group). These kinds of records became important during the nineteenth century, when scientists were first establishing the existence of the 11-year sunspot cycle, and again in the twentieth century, when they were involved in proving that there have been extended periods in the past when the sunspots just didn't appear at all. (We won't go into these periods here, but anyone interested can look up the "Maunder Minimum" to learn about the most recent of these events.)

It appears that the Chinese, in addition to keeping astronomical records, developed theories about how the heavens work—what we would call today cosmological models. Oddly enough, although there is some scholarly conflict on this point, it appears that the Babylonians may not have done this. I picture Babylonian astronomers poring over their massive collection of data looking for regularities, much as a modern stockbroker might look over stock prices for hints about future activity. The stockbroker really doesn't care if the company whose stock he is considering builds widgets or thingamabobs—all he wants to know is whether the stock price will go up. In the same way, a Babylonian astronomer would want to be able to find regularities in celestial events, and wouldn't care much about what caused those regularities. With enough data (and enough persistence) things like the occurrence of eclipses and the appearance of the new moon (the start of the new lunar month) could be teased out from the massive database they had accumulated. They could (and did) extract information without any need to understand what the moon is or why it behaves as it does.

As it happens, the Babylonian records played a major role in later scientific developments. They enabled the Alexandrian astronomer Hipparchos to determine the length of the lunar month to an accuracy of seconds, for example, and played a role in the calculations of Claudius Ptolemy, who devised what is arguably the most successful cosmological model the world has ever seen (at least if you measure success by longevity). But that is a story for another chapter.

Chapter 3

Counting

There is probably nothing about science that causes more grief to students and instructors alike than the fact that so much of it involves mathematics. In the words of Galileo, "The book of nature is written in mathematics." The use of mathematics in science arises for two important reasons. First, most sciences try to give quantitative descriptions of nature. If you want to say where a particular star will be in the sky at 9:00 tonight, for example, you have to give two numbers to tell someone where to look (the numbers are the angle above the horizon and the angle away from north). Even a simple operation like this, then, involves the use of numbers. The second reason mathematics enters science is that, as we shall see in subsequent chapters, it is often convenient to express scientific results in terms of equations rather than words. Before discussing the origins of modern mathematics in this chapter, though, I have to make one point: there is nothing written in the language of mathematics—no equation, no theory—that cannot also be written, albeit less elegantly, in plain English.

The development of mathematics began with the simple need for counting. If you are selling five bushels of wheat to someone, for example, there is an obvious need to be able to record the number "5" on whatever document accompanies the transition. You might, for example, carve five notches on to a stick. Undoubtedly, the first beginnings of mathematics started with simple counting like this. It's also not hard to see that as the number of bushels increased, you might get tired of carving notches. You might, in fact, invent a different kind of notch to stand for five or ten simple notches. Most early numerical systems had this kind of cumulative nature. We'll use Roman numerals as an example, even though they were developed relatively late in history, because they are familiar to us today.

In this system, the symbol "I" stands for one "notch on the stick." If you want to write two or three, you just add more "notches"—II and III, respectively. You keep accumulating this symbol until you get to five, which you write as "V" before you start accumulating again. VI is six, VII seven, and so on until we get to ten, which is written as an "X." The system goes on to

"L" for fifty, "C" for a hundred and "M" for a thousand. If a number appears in front of one of these symbols, it is interpreted as representing one less, so that IV stands for four and IX for nine, for example.

Roman numerals give us a perfectly fine way of counting up to quite large numbers, and works well provided that counting is all you want to do. For any more complex operations (such as multiplication and division), though, they are pretty awful. If you don't believe me, try multiplying CVII by XXV! In fact, we use Roman numerals primarily in a ritual way today, to give a kind of ersatz grandeur to some event like a Super Bowl or the sequel to a movie. In the late nineteenth century, Roman numerals were often inscribed on public buildings to mark their date of construction—MCCMLXXXVI for 1886, for example. Sometimes the pseudo dignity of Roman numerals is used to satirize events, as when a sports writer reported that in Super Bowl IX there were VII fumbles and IV pass interceptions.

As we mentioned above, Roman numerals came fairly late in the development of mathematics. The real invention of mathematics took place in Egypt and Mesopotamia millennia before the Romans came on the scene. Fortunately, civilizations in both of these areas left abundant written records of their mathematical work—sometimes even what amount to mathematics textbooks—so we have a pretty good idea of how they operated.

Egypt

The Egyptians were also an intensely practical people—one author refers to them as a "nation of accountants." They kept track of all sorts of things as a matter of course. According to one legend, for example, they counted the number of enemy soldiers killed in a battle by assembling (and counting) a pile of severed penises! Obviously, any people with this sort of mindset is going to develop a number system pretty quickly. In fact, over the course of their long history, the Egyptians developed two numerical systems, similar in organization but using different symbols. Egyptian numbering will seem familiar to you because, like our own, it is a decimal system (i.e. it is based on 10).

The unification of Upper and Lower Egypt into a single entity took place around 3000 BCE, and both hieroglyphic writing and what is called the hieroglyphic number system were probably in place at that time. Certainly the great pyramid at Giza (built around 2650 BCE) could not have been completed without both of these systems.

In the hieroglyphic number system, quite similar to the Roman numerals that came later, there was a symbol for one (a line), another symbol for ten (sort of an inverted "U"), another for 100 (a coil, something like a backward "9"), yet another for 1000 (a lotus) and so on. Numbers were written by piling up the symbols—a lotus, four coils, five inverted "U," and three lines, for example, would be 1453.

Sometime around 2000–1800 BCE, a simplified notation was developed. Scholars call this the hieratic system. Like our own system, it has separate symbols for the numbers up to 9, but unlike our system, it then went on to supply separate symbols for 10, 20, 30 ... , as well as 100, 200, 300 ... and so on. This meant that scribes had to memorize a lot more symbols, but it made writing numbers a lot less tedious.

Like the Babylonians, the Egyptians made extensive use of tables to carry out mathematical operations, with the Egyptians making extensive use of the arithmetic operation of doubling. For example, the problem $2 \times 5 = 10$ would be worked by an Egyptian as follows:

$1 \times 2 = 2$
$2 \times 2 = 4$
$2 \times 4 = 8$

So

$5 \times 2 = (1 + 4) \times 2 = 1 \times 2 + 4 \times 2 = 2 + 8 = 10$

There are also extensive tables of fractions, which Egyptians always expressed as a sum of fraction of the form 1/n; where n is an integer. For example, the table might have an entry like

$\frac{5}{6} = \frac{1}{2} + \frac{1}{3}$

With these tables, they could perform quite complex calculations—for example, in one papyrus the student is asked to calculate how to divide seven loaves of bread among ten men, a request that requires a fairly high level of mathematical sophistication.

Our knowledge of Egyptian mathematics comes from the existence of what are, in essence, a book of worked out examples for students. The most important of these is called the "Rhind Papyrus," after A. Henry Rhind, a Scottish antiquarian who bought the manuscript in Luxor in 1858 and left it to the British Museum upon his death in 1864. I should mention that in those days the laws governing archeological finds were different than they are today, and as a result many artifacts of the ancient world wound up in European and American museums. Whether this represents an act of preservation of materials that would otherwise have been lost (as some maintain) or the looting of national treasures (as argued by others) remains a subject of debate today.

The Rhind Papyrus was written around 1650 BCE by a scribe named Ahmes, who says that he copied from a manuscript written during the XIIth dynasty, around 1800 BCE. (Note the use of Roman numerals to identify royal dynasties in Egypt.) Ahmes says the text will give the correct method of reckoning, for grasping the meaning of things and knowing everything that is, obscurities and all secrets.

He then goes on to give 87 worked out examples of various types of problems.

As with the example of the division of seven loaves among ten men mentioned above, many of the examples in the papyrus involve what we would call fractions today. I would, however, like to concentrate on the subset of the problems that deal with a different subject—geometry.

The central fact of Egyptian life was the annual flooding of the Nile. Over a period of several months new topsoil was laid down over the land, ensuring the continued fertility of Egypt's farms. It was this feature of the Egyptian world, in fact, that caused the Greek historian Herodotus to refer to the country as "the gift of the Nile." It also, however, required that the Egyptians re-survey their land every year. Thus for them, geometry was not an abstract subject of intellectual interest, but a vital part of everyday life.

My favorite problem in the Rhind Papyrus is #50, which is stated as follows:

A field round of khet (diameter) 9. What is the amount (area) of it?

Take thou away ⅑ of it, namely 1. Remainder is 8. Make thou the multiplication 8×8. Becomes it 64. The amount of it, this is the area.

In modern language, Ahmes is telling us that the area of the circle is given by the formula

$$A = (\tfrac{8}{9} \times d)^2$$

whereas, if you recall your high school geometry, we know the correct formula to be

$$A = \pi \, (\tfrac{d}{2})^2$$

If you compare these two formulae, you realize that the Egyptians had actually developed an approximation for π, the ratio of the circumference of a circle to its radius. In fact, the Egyptian value of π is 3.160 compared to the correct value of 3.14159 ... —a difference of less than 1 percent. (In fact, the correct answer to problem #50 is 63.6 rather than 64.)

This exercise illustrates an important difference between the way that the Egyptians approached mathematics and the way we approach it today. To us, π is a number whose exact value matters—it's one of the fundamental constants of the universe. To the Egyptians, getting the area of a field within a margin of error of less than 1 percent was good enough. They really weren't interested in the deeper question of the nature of π, they just wanted to get on with their farming. We see this throughout their geometry. Their formula for finding the area of a four-sided field is exact only if the opposite sides of the field are parallel to each other, for example, but gave "good enough" results for most of the fields they had to deal with.

Like the Babylonians, they seem to have had some sense of the Pythagorean theorem relating the sides and hypotenuse of a right triangle. Again, you can imagine a surveyor noticing that when the two sides of triangular fields were 3 and 4 units, the third side would be 5 units long and writing this down as a general rule. Contrast this to Euclid's presentation, where the same result follows as a logical consequence of a set of axioms.

The approximate Egyptian value of π is useful for another reason. Some advocates of the notion that the pyramids were built by extraterrestrials (on the grounds, I suppose, that the Egyptians were too stupid to have done it themselves) have pointed out that if you divide the circumference of the Great Pyramid by its height, you get something close to 2π. From this they concluded that an advanced civilization had to be involved. Problem #50, I hope, lays that claim to rest. In point of fact, the Egyptians were some of the best engineers the human race has ever produced, using the materials at hand (sand and human labor) to produce monuments that have endured through the ages.

One last problem (#56) from the Rhind Papyrus to make this point.

> Example of reckoning a pyramid 360 in its ukha-thebet (length of one side) and 250 in its peremus (height). Cause thou that I know the seked of it (slope of the side).

I won't go through the solution, since the main complications have to do with the conversion of cubits (the length between elbow and fingertip—about 18") to palms (1/7 of a cubit). The end result is a calculation that tells you that the side must rise about 20" for each 18" in length. These guys knew their pyramids!

As a final note, I should mention that if you consult the modern Universal Building Code in the United States, you will see roof slopes described in exactly the same way as the slope of a pyramid was described in the Rhind Papyrus—in terms of how many feet a roof rises per how may feet in linear extent. A typical low sloped roof, for example, might rise 1 foot in height for every 4 feet of length.

Mesopotamia

Most of the written records we have from this region date back to around 2000 BCE, although there is evidence that the Sumerians had developed a number system centuries earlier. As mentioned in Chapter 2, these records were written on clay tablets with a wedge-shaped stylus, so many have survived to the present day.

The Babylonian system, the most widespread and influential numerical system in the ancient world, worked this way: To record a "one," a downward stroke of the stylus produced a "V" in the clay. To record a "ten," a sideways stroke was used—">" This means that a number that we would write as "23" would be in the Babylonian system >>VVV. Up to this point, the system seems to look a lot like Roman numerals.

It is, until you get up to "59". At this point, the Babylonian system does something with which we're familiar. It switches the numbering system over to a new column (the way we do when we write "10" after "9") and starts the counting all over again. Thus, the number "61" is written "V (blank) V." The Babylonian system, in other words, was based on the number "60" the way that ours is based on "10." The number "V (blank)V (blank) V," for example, should be interpreted as meaning 3600 (this is 60 × 60) for the first "V," "60" for the second "V" and "1" for the final "V"—the number we would write as 3661. Thus, the Babylonian system is a strange mixture of a decimal system like the one we use and a system based on 60.

Why 60? The short answer is that no one is really sure. Some have suggested that it is because the year has approximately 360 days, others that 60 is a number easily divisible by many other numbers. There is even a semi-serious suggestion that it has to do with the fact that there are 15 finger joints on the human hand (remember 4 × 15 = 60). Frankly, none of these explanations seems convincing to me. The Babylonian system does, however, point to an important fact about arithmetic: there is nothing sacred or special about number systems based on "10," the kind of decimal system we're used to.

In fact, even though the decimal system seems "natural" to creatures with ten fingers, we often use other systems (albeit unconsciously) in our daily lives. Your laptop computer, for example, uses a binary system—one based on "2." In this system the numbers are 1 (one), 10 (two), 11 (three), 100 (four) and so on. The reason this system is "natural" for computers is that the basic working unit of the machine is a transistor, which can either be on (1) or off (0). In the 1970s, computers often used a numbering system based on 16. The somewhat obscure reason was that transistors came in racks of eight, and you sometimes needed two racks (16 transistors) for calculations. In this system, the numbers ran up to 9, followed by A, B, C ... F before you got to "10." I was amazed at how quickly I and other physicists adapted to this unfamiliar system.

In any case, there are still vestiges of the Babylonian system in our lives. We divide the hour into 60 minutes, the minute into 60 seconds, and measure angles in degrees, of which there are 360 in a full circle.

Having said this, we have to point out that there is one serious deficiency in the Babylonian number system, and that is the absence of the number "zero." We'll discuss zero in more detail below, but for the moment note that without a zero to hold a place, the Babylonian one ("V") and the Babylonian sixty ("V") are indistinguishable from each other. In our Arabic system, it is easy to see that 210 and 21 are different numbers, but this isn't the case for the Babylonians. (Incidentally, the lack of a zero in the system explains why I jumped from 59 to 61 in the discussion of Babylonian numbering above.)

We mentioned in Chapter 2 that the Babylonians were really good at calculation, even though their methods may seem a little strange to us. Let's take a simple operation like multiplication. The Babylonians had large tables of the squares of numbers—$2^2 = 4$, $3^2 = 9$, $4^2 = 16$, and so on. They would then multiply numbers together by using the simple algebraic formula:

$$ab = ½ ((a+b)^2 - a^2 - b^2).$$

Let's use the Babylonian method to do a simple exercise, which we would write as

$$2 \times 5 = 10.$$

The Babylonian mathematician would say that $a = 2$ and $b = 5$, then consult his tables to find that

$$a^2 = 4$$
$$b^2 = 25$$
$$(a + b)^2 = 7^2 = 49.$$

He would then say that

$$2 \times 5 = ½ (49 - 25 - 4) = ½ (20) = 10.$$

It may seem cumbersome, but the system worked. The Babylonians also had a way to solve certain kinds of quadratic equation, a subject to which we'll return in the next chapter, and find an approximate value for the square root of 2—another difficult problem. They didn't do so well in estimating π, the ratio of a circle's diameter to its circumference, though. They approximated it by 3, rather than 3.14159 ... we know it to be.

One thing that comes through when you look through the Babylonian mathematics texts is the intense practicality of the kinds of problems they felt they had to solve. In fact, reading these texts may remind you of the "word problems" you encountered back in high school. A typical problem, for example, might tell you how much barley you had to give a man each day while he dug a canal of a certain length, then ask you to figure out how much barley you would need to hire ten men to dig a bigger canal. (Irrigation canals were a major technological advance that allowed Mesopotamian civilizations to flourish.)

Mayan

We can't leave the subject of the development of mathematical systems without mentioning the numerical system of the Mayan civilization in what is now Mexico and Guatemala. The ancestors of the Mayan people came to Central America around 2000 BCE, and the Classical Period of Mayan civilization lasted from about 250 to 900CE. Much of the records of the civilization were destroyed by the Spanish Conquistadors in the sixteenth century, but enough material has survived to give us a good sense of their mathematics.

This system developed independently of the Babylonian and Egyptian systems discussed above, but bears a striking resemblance to them. There were three symbols—a dot (for one), a bar (for 5) and a figure like a clamshell or squashed football for zero. (The fact that the Mayans had developed the concept of "zero" is interesting, but because of their geographical isolation from European and Asian cultures, this discovery did not spread.)

Counting in Mayan proceeded in a manner similar to Roman numerals—a single dot for "one," two dots for "two" and so on. The bar replaced five dots and the process repeated. The system was based on the number 20, rather than ten. For the record, this is called a "vigesimal" system, and scholars suggest that this is the result of counting on both fingers and toes. Thus, three bars under four dots would represent 19, while a dot followed by a clamshell would be 20, two dots separated by a space 21, and so on. There is a curious glitch in the system that actually may be a hint as to its origin. The next place, which you would expect to represent 400 (20 × 20) is actually 360, after which the normal progression of base 20 resumes. As far as we can tell, the Mayans, like the Romans, used their number system for counting only, and did not perform operations like multiplication or division, nor did they have fractions.

The best guess as to the origin of the Mayan number system is that it arises from work on the calendar—you will recall from Chapter 2 that the Mayans had a sophisticated knowledge of astronomy. They actually had two calendars—a religious calendar of 260 days (13 months of 20 days each) and a civil calendar of 365 days (18 months of 20 days each, and a short month of five days). Unlike the Egyptians, who used the "extra" five days for celebration, the Mayans considered this an unlucky period. Although the origin of the 260-day "year" is unclear, it has been suggested that it might correspond to the period of time that separates days when the sun is directly overhead in the tropics. (Remember that the sun is never directly overhead in the northern hemisphere.)

The Mayans had major celebrations when the two calendar cycles came into coincidence (every 52 years) and when the two came into coincidence with the appearances of the planet Venus (every 104 years). For ceremonial purposes, such as carving the date when a major monument was completed, they used what is called the "Long Count," which is basically a representation of the number of days since the creation of the world. There is some scholarly debate over the issue of when the Mayans believed this event occurred, but generally some date around August 12, 3113 BCE is suggested. Counting forward from this date, Dec. 22, 2012 will be 12.19.19.17.19 in the Mayan numerical system, and, of course, Dec. 23, 2012 will be 13.0.0.0.0. In effect, the "speedometer" (what is called a milometer in Britain) will turn over on that date.

Once you understand the meaning of the Long Count dates, you realize that Dec. 22, 2012 has no more real significance than the fact that the year changes at midnight on Dec. 31 every year. The world didn't end on last New Year's day, or on Dec. 31, 1999, and, despite what Hollywood and the blogosphere would have you believe, it won't end on Dec. 22, 2012 either!

The history of zero

In our discussion of the development of various mathematical systems, we have seen over and over the importance of the concept of zero. In some cases, such as the use of the concept by the Mayans, geographical isolation obviously restricted the ability of other cultures to adopt the innovation. This isn't the whole story, though, because over the millennia various mathematicians seem to have developed a concept like zero, only to have it drop from sight again. It would make a better story if we could say something like "On such and such a date, mathematician X discovered zero and everyone else adopted it," but it just didn't work that way.

To understand this, you have to think about how some of the people we've talked about so far might have reacted if you gave them an argument about the importance of zero. An intensely practical Egyptian, presented with a problem in which a farmer had zero cows, would probably ask "Why are you bothering me with this?" To him, if there were no cows there was no problem to be solved, and he would turn his attention to matters of more immediate concern.

A Greek mathematician, on the other hand, would (characteristically) be more concerned with the philosophical implications of the concept. "How can something (a symbol) stand for nothing?" Not an easy question to answer. You may recall that Aristotle famously argued that a vacuum (nothing) could not exist, because a void would be filled quickly by the surrounding material. His dictum, "Nature abhors a vacuum" is still quoted in fields as far apart as physics and political science. (For the record, Aristotle's argument only holds if the vacuum is not in a chamber from which surrounding matter is excluded.)

Both of these examples illustrate the difficulty inherent in abstracting the notion of number—in going from "three cows" to "three entities" to the concept of "three" in the abstract. The problem, of course, is much worse for the concept of zero.

The Babylonians, with their sexagesimal system, really needed a way of representing a placeholder like zero. They customarily used a blank space or, later, special signs like " to separate numbers in different columns. They never, however, used these marks at the end of a number, which means that the really couldn't distinguish, by writing alone, between numbers like 31 and 310. They seemed to rely on context for this. A modern example of relying on context might be the fact that if you asked how much something costs, you would interpret an answer like "two fifty" differently if you were buying a magazine or buying an expensive item of clothing.

Despite the philosophical reservations of the Greeks, it appears that the Ptolmaic astronomers actually did use zero when they did calculations. They used the Greek letter omicron (O) as a placeholder in their numerical representations, a device that made sense, since they used alphabetical letters for the rest of the numbers as well. In the *Almagest*, for example, Ptolemy used the Babylonian system with omicrons inserted both in the middle of numbers and at the end. It appears, though, that this device never left the technical world of the observatory to move into the everyday world of the Hellenistic period.

In fact, the origins of our current number system did not originate in either the classical world or Mesopotamia, but in India. I should warn you that there is considerable disagreement among historians about the exact dates and sequence of developments here. Part of this is caused by the fact that the mathematical documents we have today are not originals, but copies of copies of copies, which, of course, raises the question of what was in the text when it was written originally and what was inserted later.

The consensus narrative, however, seems to indicate that around 500 CE the Indian mathematician Arybhatta (476–550) was using a positional number system. He didn't use a symbol for zero, but he used the word "kha" to denote position. There is, incidentally, a suggestion that he also may have realized that the number π was irrational (that is, that it cannot be represented by a fraction with whole numbers on the top and bottom).

By 628 the mathematician Brahmagupta (598–660) was not only using a symbol for zero (a dot or a circle), but stated a series of rules for using zero in calculations—rules like "zero added to zero yields zero." It is clear that by this time Indian mathematicians had succeeded in making the difficult transition from zero as the absence of anything to zero as an abstract number. More surprisingly, they also developed the concept of a negative number and understood abstract principles like the fact that subtracting a negative number from zero yields a positive. Like many modern students, though, they had difficulty with the process of dividing by zero, and never quite got it right.

I should point out that we still sometimes wrestle with problems related to zero today. At the turn of the millennium in 2000, for example, there was a vocal group of dissenters who argued that since our system starts with the year 1, not the year zero, we shouldn't be celebrating the millennium until December 31, 2000. When I was cornered by these people at parties, my response was "I suppose that in a technical sense you may be right, but you're going to miss a helluva party!"

Chapter 4

Greek and Alexandrian science

It is common for people in any given civilization to look back to some previous time as a sort of Golden Age and say "this is where it all started." In western civilization today, we tend to look back in this way to people we refer to as "The Greeks." (Ironically, the Greeks themselves looked back in the same way at the Egyptians.) The term "Greek" tends to conjure up images of bearded sages in white togas wandering around Athens, but in fact Greek civilization included many different cultures around the eastern Mediterranean and lasted a long time—we'll be looking at a time period from 600 BCE to around 200 CE, for example. Given this long duration and widespread geography, it shouldn't be surprising that there was a lot of variety in Greek contributions to science. What is most important about them is that many of the questions that have been the staple of scientific development for the last two millennia were first asked by Greek philosophers. Furthermore, as we shall see in later chapters, their ideas were influential far beyond their own time and place.

For our purposes in tracing the development of science we will look at three different places and periods—think of them as snapshots of a more complex and continuous process. The first of these will involve philosophers in the Greek colonies of Ionia (in what is now Western Turkey) around 600 BCE, the second a group of more famous philosophers in Athens around 350 BCE, and the last a rather remarkable group of scholars in the Greek city of Alexandria, in lower Egypt, in the first few centuries CE. The main contributions we will encounter are (1) serious attempts to explain observations of nature in naturalistic terms (the Greek phrase for this procedure was "saving the appearances"), and (2) a strong reliance on the use of human reason (as opposed to religious faith) to understand the world.

Ionia

The Greek colonies in Ionia, the land along the eastern coast of the Mediterranean, were located at a commercial crossroads, where the east–west trade routes intersected those running north–south. This tended to produce people who were sophisticated and worldly, in much the same way that modern places of cultural confluence (think New York and London) do the same today. I have always had a feeling that modern Americans—brash, experimental, always on the lookout for something new—would have been at home in the Ionian cities. It shouldn't be surprising, then, that people to whom posterity has given names like "the first scientist" hailed from this region.

The general term applied to the philosophers we'll be talking about is "pre-Socratics," though it is obvious that this is not how they thought about themselves. With a few exceptions, there are no surviving texts from any of the men we'll be discussing. Instead, we are forced to reconstruct their views from quotations in later texts that did survive—a text that might say something like "philosopher X said such and such." Nevertheless, enough of these references survive to give us a fairly good idea of what they thought. For our purposes, we will look at how the pre-Socratics dealt with one of the great questions mentioned above: the one that asks "What is the world made of?" And in view of the ground we have to cover, we will restrict our discussion to the work of two men—Thales of Miletus and Democritus of Abdera.

Thales was born around 645 BCE in the port city of Miletus on the western coast of modern Turkey. He is generally considered the first person who began to seek out natural explanations for events, rather than supernatural ones, and is therefore often called the "First Scientist." There are many legends about Thales' life. Herodotus, for example, tells us that he predicted an eclipse of the sun that interrupted a battle between the Lydians, led by the legendary king Croesus, and the invading Persians. There are also legends that have Thales becoming wealthy by cornering the olive market and, to illustrate another aspect of his legend, that he fell into a ditch because be was looking up at the night sky and not paying attention to where he was going.

Whatever he was like as a person, Thales was clearly an original thinker on scientific subjects. According to Aristotle's *Metaphysics*:

> For it is necessary that there be some nature (substance) from which become other things … Thales the founder of this type of philosophy says that is water.

It's easy to see what Thales was driving at. In modern language, the most common forms of matter around us are solids, liquids, and gasses. There is one common substance that we routinely see in all these states. Water can be a solid (ice), a liquid (coming from your tap) and gas (steam). What could be more natural than to say that water is the fundamental constituent of all matter?

Followers of Thales later added Earth, Fire, and Air to his Water to produce the well known Greek quadrumvirate of elements. The important point about this theory is not its accuracy, but the fact that for the first time human beings had produced an accounting of nature that didn't depend on the whims of the gods, but on purely natural causes. This was a major step, to be sure.

The next important step was the product of a group of philosophers, the best known being Democritus. He was born about 460 BCE in the town of Abdera in Thrace, in what is now the narrow neck of land that connects Turkey and Greece. Like Thales, Democritus thought about the fundamental constituents of the universe, but unlike Thales he approached the problem from a theoretical (or philosophical) point of view.

Suppose, he asked, you took the world's sharpest knife and began cutting up a piece of material—think of a block of wood. You'd cut it in half, then cut one of the halves in half, then one of *those* halves in half, and so on, until you came to something that could no longer be divided. Democritus called this the "atom," which translates roughly as "that which cannot be divided." The Greek atomic theory was actually pretty well thought out, with the properties of different materials being related to the (hypothetical) properties of the atoms. Iron atoms were supposed to be solid chunks of stuff with hooks that bound them together, for example, while water atoms were smooth and slippery. This theory never really caught on, however, and aside from the word "atom," had little effect on the further development of science (see Chapter 7).

The Athenians

As mentioned above, when most people hear the term "ancient Greeks," they think of the Golden Age of Athens in the fourth century BCE. More explicitly, they think of the "Big Three" of Athenian philosophy—Socrates, Plato, and Aristotle. It is important to realize that although these men, particularly Aristotle, played a significant role in the development of science, they would not have considered themselves scientists in the modern sense of the term. They were philosophers, concerned with issues like understanding the nature of reality or discovering how to live the good life. In what follows, then, we will be looking at only a tiny fraction of their thought—the fraction that bears on the later development of science.

In his *Republic*, Plato presents a compelling analogy to explain his view of reality. Human beings, he said, were like prisoners in a cave, watching shadows on a wall. The real world—the things making the shadows—is outside, not available to our senses, but we can know this reality through the power of reason. The ultimate reality for Plato was things he called "forms," which embody the pure essence of being. For example, there is a form for "dog," and every dog we see is an imperfect representation of that form.

Here's an example I've found useful in thinking about forms. Think back to the day in your high school geometry class when your teacher proved the Pythagorean theorem ($a^2 + b^2 = c^2$). Chances are that he or she drew a right triangle on the board to help with the proof. Now ask yourself a simple question: exactly what triangle was your teacher talking about?

It's clearly not the triangle on the board—the lines aren't absolutely straight and it has all kinds of imperfections. Your teacher was talking about an ideal, perfect triangle—one that doesn't actually exist in our world. That perfect triangle is a form, and every real triangle is only an imperfect representation of it.

Although we no longer talk about Platonic forms except in philosophy classes, this way of looking at the world led to an extremely important notion in the sciences: the notion that we can use mathematics to describe the physical world. We'll devote the entire next chapter to this concept, but here we simply note that if you believe that you arrive at a knowledge of true reality through reason, and if you believe, as the Greeks did, that geometry is the highest form of reason, then it's a very short step indeed to the notion that the world can be described in mathematical terms.

Perhaps the best known application of this geometrical notion was made by a sometime student of Plato's named Eudoxus of Cnidus (about 410–355 BCE). Plato is supposed to have asked his students to find the simple, regular motions in the heavens that would explain the irregularities of the observed motion of the planets—an activity he called "saving the appearances." (Today, of course, we realize that most of the apparent irregularities arise because we are observing the solar system from a moving platform.) The idea that the Earth has to be the center of the cosmos was so ingrained, however, that this solution occurred to only a few Greek philosophers over the centuries. Consequently, in the model Eudoxus developed, the Earth sat at the center, unmoving. The sun, the moon, and the planets were each moved by a family of spheres that traced out the motion of the body through the sky. This model actually explained a good deal of what astronomers had observed, although it ultimately failed to explain why the apparent brightness of planets changes over time—an effect we now know is caused by the fact that the Earth moves in orbit around the sun.

It was Plato's student Aristotle (384–322 BCE) who had the greatest impact on the development of science. He had an adventurous life, serving as a tutor to Alexander the Great and founding his own school (the Lyceum) in Athens. He wrote on a wide range of topics, including many that we would today consider scientific. Unfortunately, although he is supposed to have written around 150 books, only some 30-odd survive today.

Reading Aristotle can be a little difficult. Whereas the dialogues of Plato are polished literary works, what we have of Aristotle tends to be choppy and broken up into short segments, more like lecture notes than a finished product. Nevertheless, the power of his intellect was such that he defined the way that science would proceed for over a millennium.

Although Aristotle wrote on many phases of science, he comes across to the modern reader as a biologist—in fact, his studies of ecosystems in the eastern

Aegean are a really good example of field ecology. He rejected Plato's ideas of the forms in favor of studying nature directly. His method of analysis consisted of identifying four "causes" of everything in nature. (A word of caution: the word "cause" in the Aristotelian sense isn't the same as what we're used to in modern English.) The "causes" are:

Efficient cause—what made this thing
Formal cause—what is its shape
Material cause—what is it made of
Final cause—why was it made.

For example, if you analyzed a boat in this way, the efficient cause of the boat would be the designer and builders, the formal cause the shape of the boat, the material cause the wood from which it was made, and the final cause the task of moving goods over the water. The idea is that by examining everything in nature in this way, we can produce a codified, interconnected body of knowledge about the world.

What Aristotle gave the future, then, was a coherent and logical way of examining nature. His emphasis on "causes," particularly final causes, dominated natural philosophy for the next millennium and a half. Sometimes, however, this emphasis turned out to set up barriers to scientific advancement. We can look at one example—the problem of what is called projectile motion—to see how that could happen.

The central notion of Aristotelian physics is that the behavior of objects is determined by their inner nature. A heavy material object falls, for example, because in some sense its nature compels it to seek out the center of the universe (which, remember, for the Greeks was the same as the center of the Earth). In Aristotelian terms, the final cause of a rock is to fall to the center. This is why the rock falls when you let it go. Aristotle called this fall the rock's "natural motion."

If you throw the rock up, however, you are imposing a motion on the rock against its nature. This is called "violent motion." By setting up these two categories of motion, Aristotle posed what is basically an unanswerable question: if you throw the rock, when does the natural motion take over from the violent? You can still find evidence of medieval philosophers wrestling with this question centuries after it was first posed.

It wasn't, in fact, until the work of Galileo in the seventeenth century (see Chapter 6) that the problem of projectile motion was solved, and the solution looked nothing like what Aristotle might have imagined. In effect, what Galileo said was that he could tell you anything you wanted to know about the rock— where it will be at any time, where it will land, how long it will be in the air. He cannot, however, tell you about natural and violent motion, simply because those categories have no meaning for the problem. We now realize that many of the categories Aristotle imposed on nature come from the human mind, not from nature itself, and are therefore best ignored.

Alexandria and the Hellenistic Age

The world of classical Greece is generally reckoned to end with the massive conquests of Alexander the Great (356–323 BCE). It is one of those strange aspects of the American educational system that, while students are often exposed to the philosophers and culture of classical Athens, the period immediately following it is ignored. It's almost as if history skipped from Alexander to Julius Caesar without ever actually passing through the several hundred years in between. In point of fact, this period, which historians call the "Hellenistic" era, was an extremely important time in the development of science, as we shall see. This was, after all, the period in which Archimedes of Syracuse (287–212 BCE) discovered the principle of buoyancy, among other things; Aristarchus of Samos (310–230 BCE) first suggested that the sun was at the center of the solar system; and Hipparchus of Rhodes (190–120 BCE) made a credible attempt to measure the distances to the moon and sun.

Some background: Alexander the Great, after unifying Greece, conquered most of the known world, reaching as far east as modern day India. Upon his death, his great empire was split up between four of his generals, and one of them, a man named Ptolemy, took the rich country of Egypt as his share. Alexander had founded over a dozen cities named "Alexandria" during his lifetime, but the city in Egypt was by far the most important. Located on the Mediterranean at one side of the Nile delta, it became a major center of learning and commerce.

Having said this, I should hasten to point out that Alexandria was not really Egyptian, but a Greek city that happened to be located in Egypt. Although Ptolemy and his successors took the ancient title of "Pharaoh," it would be two centuries before the last of the line—the famous Cleopatra—bothered to learn the Egyptian language. One writer captured the status of Greek Alexandria quite well by referring to it as a "gated community." In any case, Alexandria was ideally poised to use the wealth of Egypt to continue the advance of classical knowledge. Two institutions—the Museum and the Library—were important pieces of this effort.

The Library of Alexandria was famous in its own time and remains a legend even today. The first thing you have to realize is that it was not at all like what we would call a library today. For one thing, there were no books. Alexandria was a center for the manufacturing of papyrus, and the writing was actually done on scrolls. The walls, in other words, would look more like a set of small cubbyholes than an array of shelves. The Library was part of what we would call a research institute today, a place where scholars from all around the classical world would come to work and live, supported by the Ptolemys. It was supposedly modeled on the Lyceum, the school Aristotle founded in Athens. We don't have any drawings or architectural plans for the building, but some sources suggest that it had large reading rooms and separate offices for cataloguing and acquisitions. The Museum was even less like its modern counterpart, since it was not open to the public and had no exhibits. Instead, you can think of it as a kind of small think tank attached to the Library.

What made these institutions so successful was the support of the entire line of Ptolmaic rulers. Collecting scrolls was apparently something of a family passion. There was a law that said that every ship that entered the harbor was to be searched for books. If any were found, they were seized and copied, with the originals going to the Library and the owner getting the copy. Another story, of somewhat dubious accuracy, mentions that Mark Anthony gave Cleopatra (the last of the Ptolemys) several hundred thousand scrolls from the library at Pergamon as a wedding present. The library of Alexandria was burned down some time in antiquity. Depending on who you believe, it was destroyed as collateral damage when Julius Caesar set fire to the Egyptian fleet in Alexandria harbor in 48 BCE or at the order of a Christian bishop in 391 CE. Whichever is correct, the great store of knowledge, collected over centuries, was lost. On a happier note, a new Library of Alexandria is being built on the waterfront of the ancient city, a fitting memorial to a great institution.

Foundations like the Library and Museum, persisting over several centuries, are bound to boast many famous scholars and scientists. We will look at the lives of only three of these luminaries—Euclid (around 300 BCE) who codified the discipline of geometry; Eratosthenes of Cyrene (276–195 BCE), who measured the circumference of the Earth; and Claudius Ptolemy (around 100 CE) who constructed one of the most successful models of the cosmos ever devised. Each of these had a major impact on succeeding generations, an impact that in some cases persists to the present day.

Euclid

Chances are that when you took your high school geometry class, you were taught by a method first developed by Euclid of Alexandria. We know very little about the man himself, although we know that he lived during the reign of the first of the Ptolmaic rulers of Egypt, around 300 BCE. There is some suggestion in ancient texts that he had been a student at Plato's Academy in Athens, and a story (most likely apocryphal) that when Ptolemy I asked him if there wasn't some simple way to master mathematics, he replied that "there is no royal road to geometry." Other than a few tidbits like these, our knowledge of this man and his work rests mainly on his geometrical treatise, titled *Elements*.

It's not that Euclid only worked on geometry—far from it. Like most of the important Alexandrian scholars, he worked in many areas. Fragments of his work still survive on subjects like the theory of mirrors, optics and perspective, and conic sections, to name a few fields to which he contributed. There is no doubt, however, that it is his work on geometry that has contributed most to the advance of science over the centuries.

It's important to keep in mind that Euclid did not himself derive all of plane geometry. In fact, most of the results he derived had been known, at least in a rough form, well before him. The Egyptian need to re-survey their land after each Nile flood, for example, had given them a good working knowledge of geometry, as we saw in the next chapter. Many of the ancient civilizations seemed to have some working knowledge of the Pythagorean theorem, or at least of the 3-4-5 right triangle. What Euclid did was to show that all of these disparate results could be put together in a simple logical system, starting with what he called "postulates" and "common notions" and what we usually call axioms and definitions. From this small number of initial assumptions, he developed the method of formal proof which showed that all of the rest of geometry followed from the axioms and simple logic.

For example, his most interesting axiom says that through a point outside of a line, one and only one other line can be drawn that never intersects (i.e. is parallel to) the original line. This makes intuitive sense—if you imagine tilting that parallel line, however slightly, you'll see that it will eventually intersect the original one. From this axiom (also called the "parallel postulate") it is possible to prove that the sum of the three angles of a triangle must add up to 180 degrees, another familiar result from high school geometry. It is also possible to prove the Pythagorean theorem, not as the result of lots of measurements in a muddy field, but as a logical consequence of the postulates.

So powerful was Euclid's logic that his method found use in fields far from geometry. In Chapter 7, for example, we will discuss the work of Isaac Newton and the foundations of modern science. Anyone who reads Newton's *Principia* is immediately struck by the fact that Newton is clearly modeling his presentation of dynamics, a field that has little to do with geometry, on Euclid's *Elements*.

In fact, I would argue that in this sense Euclid's example has had a negative influence on the way that modern scientists present their work. As we saw in Chapter 1, real science is based on experiment and observation. It's messy, and scientists often spend a lot of time exploring (and ultimately rejecting) blind alleys. Read a scientific paper, however, and there is seldom a hint of this real process. Instead, you have a steady progression from initial postulates to final result, with little attention paid to the actual discovery process. Personally, I see this as the hand of Euclid still at work in our world.

We need to tackle one more subject before we leave Euclid, a subject that became very important in the early twentieth century when Albert Einstein was developing the theory of relativity. The point is that Euclid's geometry, as beautiful and logical as it is, does not really describe the world we live in.

We can see this quite simply by thinking about triangles. As we pointed out above, given Euclid's postulates, we can prove that the angles of a triangle have to add up to 180 degrees. But think about a triangle on the surface of the Earth whose two sides are meridians of longitude and whose third side is a piece of the equator. The meridians come together at the North Pole to form one angle of the triangle, but they are each perpendicular to the equator. This means that the angle formed when each meridian meets the equator is 90 degrees, and the two angles together add up to 180 degrees. This, in turn, means that when we add in the angle at the poles, the angles in this triangle will add up to more than 180 degrees.

This simple example shows that Euclidean geometry doesn't describe a surface like that of the Earth. In fact, in the nineteenth century mathematicians realized that Euclid's postulates, as logical and clear as they seem, are only valid on a flat plane. They are a good approximation to the geometry of a field in Egypt or a housing lot in America, but can't be applied to every system. In fact, surveyors routinely correct their readings for the curvature of the Earth when measuring large distances. As we shall see in Chapter 9, the development of what are called non-Euclidean geometries was extremely important in twentieth-century physics.

Eratosthenes

Despite what used to be taught in elementary schools, the idea that the Earth is round has been with us since antiquity—by the time Columbus sailed, it had been accepted for millennia. In fact, the first recorded measurement of the diameter of the Earth was done about 240 BCE by the Greek geographer Eratosthenes, who later became chief librarian of the Library of Alexandria. His method worked this way: He knew that on the summer solstice light from the sun penetrated all the way to the bottom of a well in the modern city of Aswan, indicating that the sun was directly overhead. At the same time he measured the length of a shadow cast by a pole of known height in Alexandria. From this measurement, supplemented by some simple geometry, he concluded that the distance between Alexandria and Aswan was ¹⁄₅₀ of the circumference of the Earth. How he determined the distance between these two cities remains one of those amiable scholarly mysteries that will probably never be resolved, but he reported a circumference of 252,000 "stadia," where the "stade" was a standard unit of length in his time.

The problem is that there were several definitions of the "stade" in the ancient world, just as we have the statute mile (5,280 feet) and the nautical mile (6,076 feet) today. All of the stadia were around 200 yards long—twice the length of a football field—and the most probable choice has Eratosthenes' measurement of the Earth's circumference at about 29,000 miles, compared to its current value, which is just under 25,000 miles.

Not bad!

Claudius Ptolemy

Claudius Ptolemy (about 90–168 CE) lived in Alexandria at a time when it was no longer the capital of an independent nation, but the main city in one more province of the Roman Empire. His first name, the same as that of one of the Roman emperors, indicates that he was a Roman citizen. He wrote in Greek, as did all the scholars in Alexandria, which probably indicates that he was part of the Greek community in Egypt. During the Middle Ages, he was often portrayed as a king by artists, but there is no evidence that he had any relation to the Ptolmaic pharaohs. In fact, "Ptolemy" was a fairly common family name in the Macedonian nobility; there were a number of officers of that name in Alexander the Great's army, for example.

As was the case with Euclid, we know very little about the details of Ptolemy's life. Like Euclid, he worked in many fields—geography, optics, and music theory, to name a few—but his memory survives today because of one important book he wrote. A comprehensive model of the solar system, it was originally titled "Mathematical Treatise," but comes to us today as the *Almagest*. The name is a combination of the Arabic prefix and the Greek root for "great." The strange combination tells us something of the history of the work, which was lost to the west during the middle ages, then re-introduced to Europe from the Arabic world in the twelfth century.

If we judge a theory by the length of time it dominates the thoughts of educated people, then the *Almagest* contains, quite simply, the most successful scientific theory ever propounded. From the time it was written in the second century, it was the accepted picture of what people then called the universe (and we today call the solar system) until well into the seventeenth century both in Europe and the Middle East. Let's take a look at the universe that Ptolemy constructed to try to understand its staying power.

Like all Greek cosmologies, the *Almagest* starts with two basic, unquestioned assumptions: that the Earth is the unmoving center of the universe, and that the motions of the heavenly bodies have to be described in terms of circles and spheres. The first of these we've already seen in many ancient astronomies, and can clearly be attributed to the clear evidence of the senses. The second is less obvious to us, but is related to the notion that the circle is the most "perfect" of geometrical figures, and that the heavens, being pure and unchanging, must reflect this fact. The notion of the "perfection" of the circle comes easily into our minds—try asking your friends what the most perfect shape is if you don't believe me. The problem with this idea is that the notion of "perfection" really isn't very well defined—exactly what does it mean to say that a geometrical figure is "perfect," anyway? In any case, this notion was firmly rooted in ancient thought.

Ptolemy actually made very few astronomical observations himself, but relied on centuries of data taken by Babylonian and Greek astronomers, particularly Hipparcos. It was from this compilation that he put together the model we'll describe below. Although the system may seem overly complex to us, you have to remember that Ptolemy was trying to describe the motions of planets that actually move in elliptical orbits as seen from a planet that is also moving. Given the realities of planetary motion and Ptolemy's unchallenged assumptions, it's not hard to see why the system turned out to be so complex.

The book has many sections—a star catalogue and a list of constellations, for example. My own favorite is a short section in Book One titled "That Also the Earth, Taken as a Whole, is sensibly Spherical." In this section Ptolemy presents a number of observations that prove his point—the fact that a ship's hull disappears first when it sails away from land, the fact that eclipses occur at different times at different locations, the fact that during a lunar eclipse the Earth's shadow on the moon is curved. When I lecture on this topic, I like to add a modern entry, which I call the "Rose Bowl Proof." If you live on the east coast of the United States, it is dark by the time the Rose Bowl game starts in California, yet the sun is still shining over the stadium. If the Earth were flat, the sun would have to set at the same time everywhere, which it clearly does not.

The central feature of Ptolemy's model of the universe is something called the epicycle. The easiest way to picture the way an epicycle works is to imagine a large glass sphere spinning on its axis. Ptolemy called this sphere the "deferent." Inside this large sphere, imagine a small sphere rolling around, and imagine that a planet is attached to the surface of that small sphere. It is this smaller sphere that is called the "epicycle." To an observer standing at the center of the large sphere, the motion of the planet will not appear to be uniform, but will speed up and slow down at a rate depending on how fast the two spheres are rotating. Thus, even though the only motions in the system are the uniform rotations of perfect spheres, the motion of the planet can be irregular.

It gets worse. Just using epicycles wasn't enough to make the observed motion of the planets match his model—after all, the planets actually move in elliptical orbits, not combinations of circles. To deal with this sort of complication, Ptolemy had to say that each deferent rotated around a point away from the Earth called the "eccentric," and that the sphere would appear to move uniformly if viewed from a point opposite the Earth from the eccentric called the "equant." Thus, although our home planet didn't move, it was not really the center of motion of the universe.

The system was complicated, but by adjusting the size and rotation rates of all of the spheres and epicycles, Ptolemy was able to make it match the data. Furthermore, although using the system required a lot of calculations, those calculations were fairly straightforward. Thus, astronomers could use the system to predict things like eclipses without having to worry too much about its theoretical underpinnings. Later, some Arab astronomers would try to refine the system by adding still more epicycles (spheres-within-spheres-within-spheres), but the essential complication remained a defining hallmark of Ptolemy's work. Even today, a scientist who feels that a theory has become unnecessarily complicated may accuse a colleague of "just adding epicycles."

Perhaps the best commentary on Ptolemy was delivered by Alfonso the Wise (1221–84), king of Castile. Upon being introduced to the *Almagest*, newly translated from the Arabic, he is supposed to have remarked: "If the Good Lord had consulted me before the Creation, I would have recommended something simpler."

Islamic science

The Hellenistic science developed in Alexandria marked the high tide of the science of Classical Antiquity, for the storm clouds were already gathering over the Roman Empire, or at least that part of it that circled the Mediterranean. Those who like to assign precise dates to historical events mark September 4, 476 as the "fall" of the Roman Empire. In fact, this was the date when the last of the western Roman Emperors, the inappropriately named Romulus Augustus, was deposed by a Germanic general.

But this is questionable at best. In another of those mysterious gaps in our educational system, people who talk about the "fall" of the Roman Empire in terms of events in Italy ignore the fact that the other half of the Roman Empire, centered in Constantinople, was scarcely affected by the barbarian invasions. In fact, Greek speaking people who considered themselves to be Romans continued to rule in that city for another thousand years, until the final conquest by the Turks in 1453. In addition, the Germanic tribes that invaded Italy often considered themselves the inheritors and preservers of the Roman tradition, not the destroyers of a great civilization. The decline of the Roman Empire was a slow affair, and historians still argue over which of the many possible causes was most important. At best, events in 476 should be seen as symbolic of a centuries-long process of dissolution in the western Mediterranean.

By the seventh century, events in a couple of sleepy towns on the Arabian peninsula changed the entire picture of human political (and scientific) history. The Prophet Mohammed (570–632) founded the religion of Islam and after his death armies swept out of Arabia to conquer much of the Middle East and North Africa. By the end of the eighth century, after what has to be one of the most astonishing sustained military campaigns in history, Muslim armies had carried their religion from India to Spain. By conquering North Africa, they cut Europe off from its traditional agricultural supplies and instituted what historian Henri Pirenne called a "blockade" of that continent. These conquests started what scholars call the "Golden Age" of Islam, which lasted roughly until the thirteenth century. In another symbolic event, Mongol armies sacked

Baghdad in 1258, and this is often taken as the end of the Golden Age. From the point of view of the development of science, the center of gravity during this period shifted away from the eastern Mediterranean basin to the centers of Islamic culture in the Middle East and Spain.

A word about nomenclature: terms like "Islamic science" and "Islamic scholarship" can be a bit misleading. As they are customarily used, they refer to events that took place in parts of the world under Muslim rule. They should not be taken to imply that the advances we'll be discussing are in some sense an outgrowth of the Muslim faith, any more than later events we'll be describing in Europe could be termed "Christian science." Nor should the terms be understood to mean that the people involved were necessarily Muslims. Many were, of course, but the Islamic empires tended to be tolerant of other religions, and many of the scholars involved in the advances we'll be describing are actually Jewish, Christian, or followers of many of the other faiths practiced in those regions.

Having said this, we also have to note that religious institutions are an important part of any culture, so that religious attitudes and teachings will inevitably play some role in the development of science. In the next chapter, for example, we will discuss the argument put forward by some scholars that the rise of Protestantism in northern Europe, with its emphasis on individualism, was important in the development of modern science. In this chapter, we will see how Islamic attitudes toward the natural world facilitated the advance of medicine. Early Islamic writings, particularly the Hadith (writings concerning the life of Mohammed) were particularly important in this respect. Sayings like "There is no disease which Allah has created, except that He has also created its treatment," attributed to the Prophet, were obvious encouragements for the rapid advances in medicine during this period.

For our purposes, we have to note two important consequences of the Islamic Golden Age on the history of science. One, to which we have already alluded, was the preservation of much of the learning of classical antiquity. Ptolemy's *Almagest*, for example was translated into Arabic early on, and when it was translated into Latin it became a foundational document for the development of European astronomy. Many of the works of Aristotle became available in the same way.

According to legend, the Muslim interest in Classical Antiquity began with the Caliph al-Mamum (809–33), who is supposed to have had a dream in which Aristotle appeared and talked to him. Upon awakening, the story goes, the Caliph issued an order to collect and translate Greek texts into Arabic. Regardless of whether the Caliph actually had this dream, we do know that for centuries this process of translation and preservation went on. In fact, it was not at all unusual for peace treaties between the Caliphs and the Byzantine Emperor to include the requirement that books from libraries in Constantinople be copied and transferred to Baghdad.

But it was not only as conservators of previously developed ideas that Islamic scholars drove the development of science during the Golden Age. As you might expect in an empire as widespread, wealthy, and long lasting as this one, scientists in the Islamic world made their own advances in a wide variety of fields, from optics to anatomy to logic. No less an authority than the *Guinness Book of World Records* recognizes the University of Al Karaoine in Fez, Morocco, as the world's oldest functioning university (it was founded in 859), with Al-Azhar University in Cairo (founded 975) not far behind. Both of these institutions, along with many schools that issued medical diplomas, flourished during the Islamic Golden Age. A general characteristic of Islamic science was a much greater reliance on experimentation and observation than we have seen up to this point. This is a clear contrast with developments in Greece, where the emphasis was much more of reason and logic.

Confronted with a wide array of topics in Islamic science, we have to make some choices. In the end, we will concentrate on only three fields—mathematics, medicine, and astronomy—as representative of a much wider breadth of learning. Before we move into these subjects, however, let me make one observation: whenever you encounter a word that begins with the Arabic article "al," such as algebra, algorithm, or Aldebaran, the chances are that you are looking at a product of the Golden Age.

There is one more general point to be made about Islamic science. None of the great scholars we will be discussing confined their work to just one discipline. To use the term in general usage, they were polymaths. We will be forced in what follows to concentrate on the aspect of their work that had the greatest lasting impact, but keep in mind that a man now honored as a mathematician probably worked in astronomy, medicine, and philosophy, and may have written poetry on the side as well.

Islamic mathematics

In previous chapters we traced the evolution of number systems, the invention of zero and "Arabic" numerals in India, and the development of Euclidean geometry in Alexandria. These formed the basis for the advances into higher mathematics that occurred in the Islamic world.

As we pointed out earlier, there are obvious practical and commercial reasons for the development of mathematics. In the case of the Islamic world, scholars point to three other, less obvious, motivations. One was the Islamic laws of inheritance, which drove the development of the mathematics of fractions. In fact, our current practice of writing fractions as a numerator over a denominator, with a bar separating the two, was developed by a Moroccan mathematician named Al-Hassar who specialized in inheritance law in the twelfth century.

The second motivation concerns the calendar. In Chapter 2 we pointed out that nature presents us with a number of "clocks," with the day and the year being the most obvious. We also pointed out that there is a third "clock" given by the phases of the moon. These three clocks tick at different rates, and we discussed the construction of Stonehenge as a way of reconciling the first two.

Muslim societies used the third "clock," the so-called lunar calendar. In their system each month started on the day when the sliver of the new moon is first visible in the evening sky. Since religious holidays were specified according to the lunar calendar in both the Muslim and Jewish traditions, predicting the occurrence of the new moon became an important task for astronomers and mathematicians. (Incidentally, the fact that Easter occurs at a different date every year results from the fact that it is tied to the Jewish celebration of Passover, which is determined by the lunar calendar.)

Predicting the time of the new moon is actually a pretty tricky process, even for skilled mathematicians who had access to the *Almagest*. The problem is that all of Ptolemy's spheres and epicycles are defined (in modern language) relative to the plane in which the orbits of the planets lie—the so-called "ecliptic." A new moon, on the other hand, has to be defined in terms of the observer's horizon, and this depends on the latitude and, to a lesser extent, the altitude of the observatory. The need to reconcile these two frames of reference was a powerful motivation for the development of Islamic computational techniques.

Finally, although the third motivation may seem mundane, it is actually quite difficult. Muslims are supposed to face Mecca when they pray, and this means that it is necessary to determine the direction to that city from the site of every mosque in the empire. This turns out to be a rather difficult problem in spherical geometry, and accounts for a good deal of Islamic work in that field.

We have already discussed one aspect of Islamic mathematics—the so-called "Arabic" numerals. These were developed in India, as pointed out in Chapter 3, and were called "Hindu Numerals" by Islamic writers. Islamic mathematicians introduced a number of refinements such as the fractional notation mentioned above, and passed the system on to Europe. The most influential book involved in that transmission was titled *Liber Abaci* ("The Book of Calculation"), which was published in 1202 by the Italian mathematician Leonardo of Pisa (1170–1250), better known by his nickname Fibonacci.

In examining the story of Arabic numerals we encounter a phenomenon that recurs often in discussions of Islamic science. We know that the Persian astronomer Jamshid al-Khazi used the modern decimal point notation in the fifteenth century. We also know that this notation was used earlier by Chinese mathematicians and was introduced to Europe by the Flemish mathematician Simon Stevin (1548–1620), yet there is no evidence that any of these men was aware of the other's work. The laws of mathematics follow from simple logic and are accessible to anyone, so we shouldn't be surprised if different scholars discover the same things independently. We shall see this same process of independent discovery later on when we talk about the laws of nature that make up the disciplines we call science.

The most important development in Islamic mathematics, however, was not the refinement of Arabic numerals, but the development of algebra. This advance can be traced clearly to a single man—Abu Abdallah Muhammad ibn Musa al-Khwarizmi (780–850)—and a single book, titled *Kitab al-Jabr wa-l-Muqubala* ("The Compendious Book on Calculations by Completion and Balancing"), published in 830. Our word "algebra" comes from the Latinization of *"al-Jabr"* (technically, from the title of a twelfth-century translation of the book titled *Liber algebrae et almucabala*). Our word "algorithm," referring to a procedure for solving problems, also comes from a Latinization of al-Khwarizmi's name.

As often happens with early scientists, we have very little definitive information about the early life of al-Khwarizmi. He was a Persian, most likely born in the region of modern day Uzbekistan. Some scholars have suggested that he was a follower of the Zoroastrian religion, one of many faiths commonly practiced in Persia, even though his book has a preface that suggests that he was a devout Muslim.

In any case, we do know that he traveled to Baghdad, where he was affiliated with an institution known as the "House of Wisdom." Some explanation: in 762 the Caliph al-Mansur (714–75) founded the modern city of Baghdad, moving the capital of the Islamic Empire from Damascus. Following the tradition of the conquered Persian Empire, he founded the House of Wisdom as a center of learning and translation affiliated with the royal household. At first, the institute was centered on translating Persian works into Arabic. (Arabic played much the same role in the Islamic empire that Latin did in Christian Europe, serving as a kind of *lingua franca* shared by scholars of many different linguistic backgrounds.) Later, the caliph's son al-Mamum (809–33), to whom we have already referred, shifted the focus to Greek works and brought in astronomers and mathematicians, among others, to expand the work beyond just translation.

The basic contribution of Islamic mathematics is that it brought the discipline to a new, more formal state, rather different from the Greek concentration on geometry. To see what this means, let me give you an example of what the words "balancing" and "completion" mean in *al-Jabr*. I'll do this in modern language and notation, because the original text, written in ordinary words rather than symbols, is hard for present day readers to comprehend. Let's start with the equation

$$5x^2 - 10x + 3 = 2x^2 + 1.$$

The basic concept of "completion" involves performing the same operation on both sides of this equation in order to group similar expressions together.

When I teach this concept, I call it the "Golden Rule" of algebra: Whatsoever thou doest unto one side of the equation, do thou also to the other. In this case, we would begin the process for our example by subtracting $2x^2$ from each side of the above equation, yielding

$$3x^2 - 10x + 3 = 1.$$

"Balancing" refers to the process of getting terms of the same type on the same side of the equation, in this case by adding $10x - 3$ to each side, yielding

$$3x^2 = 10x - 2.$$

The problem of solving a quadratic equation (i.e. an equation including a square) had been attacked by mathematicians for millennia. The Babylonians, using the kinds of techniques discussed in Chapter 3, had been able deal with what were in essence some simple forms of these equations, and Indian mathematicians had extended that work to include more complex forms. Al-Khwarizmi, however, was the first to write down the fully general solution and to understand that such equations can have two solutions. (For example, the equation $x^2 - 2x = 0$ has solutions for $x = 0$ and $x = 2$.)

Over the centuries, Islamic mathematicians built on these sorts of foundations to construct an impressive edifice of scholarship. They recognized that quadratic equations can have negative numbers as roots, explored the realm of irrational numbers like square roots, and found the general solutions to cubic equations (i.e. equations containing x^3). They extended earlier Greek work in geometry and developed the basic functions of trigonometry (sines, cosines, and tangents). My personal favorite Islamic scholar, the Persian polymath Omar Khayyam (1048–1131) explored the boundaries of non-Euclidean geometry (see Chapter 9) when he wasn't busy writing poetry. (If you haven't read Edward Fitzgerald's translation of his *Rubaiyat*, you have a real treat in store for you.) In the end, Islamic mathematicians began searching down many paths that were later explored in more detail by other scholars in other times and places.

Pre-Islamic medicine

There is probably no more universal branch of science than medicine, and every society we've discussed up to this point made advances in this field. We will look at medicine in only three different pre-Islamic cultures—Egypt, Greece, and China—as a sampling of how the field had developed before the eighth century.

There is no question that Egypt produced the best doctors during the time when it was a power in the ancient world. We have documents attesting to the high regard in which Egyptian medicine was held—letters from kings and princes to a pharaoh, for example, asking that Egyptian doctors be sent to deal with specific problems like infertility. Doubtless the practice of mummification, which involved the removal of internal organs from cadavers, gave Egyptian physicians some knowledge of human anatomy.

As was the case for our knowledge of Egyptian mathematics, our best insight into Egyptian medicine comes from the existence of a papyrus. The so-called Edwin Smith papyrus, named after the man who bought the document in 1862 and whose daughter donated it to the New York Historical Society in 1906, is the oldest surgical text known. The papyrus was written in the sixteenth century BCE, but is based on material from at least a thousand years earlier. It has been suggested that the original text was written by the legendary Imhotep (c. 2600 BCE), architect to the pharaohs and the founder of the science of medicine.

The papyrus is basically a set of instructions for the treatment of injuries—the kind of thing an emergency room physician would read today. There are 48 cases given to illustrate treatment techniques. For example, Case 6 is as follows:

> Case Six: Instructions concerning a gaping wound in his head, penetrating to the bone, smashing his skull, (and) rending open the brain of his skull.
>   Examination: If thou examinest a man having a gaping wound in his head, penetrating to the bone, smashing his skull, (and) rending open the brain of his skull, thou shouldst palpate his wound. Shouldst thou find that smash which is in his skull [like] those corrugations which form in molten copper, (and) something therein throbbing (and) fluttering under thy fingers, like the weak place of an infant's crown before it becomes whole—when it has happened there is no throbbing (and) fluttering under thy fingers until the brain of his (the patient's) skull is rent open—(and) he discharges blood from both his nostrils, (and) he suffers with stiffness in his neck ...
>   Diagnosis: [Thou shouldst say concerning him]: "An ailment not to be treated." Treatment: Thou shouldst anoint that wound with grease. Thou shalt not bind it; thou shalt not apply two strips upon it: until thou knowest that he has reached a decisive point.

Not all the cases are this hopeless, of course. In many of them the physician is instructed to say "An ailment I will treat" and is then given instructions as to how to proceed. In fact, what we have in this very old text is something like the modern method of triage, by which physicians in an emergency decide which patients would benefit most from medical attention and act accordingly.

Two thousand years after Imhotep, a man appeared on the scene in Greece who is often called the "Father of Medicine." I am referring, of course to Hippocrates (460–370 BCE). Born on the island of Kos in the western Aegean, just off the coast of modern day Turkey, he founded a school for the training of physicians that was to have an enormous impact on medicine. He is perhaps best known today for the Hippocratic Oath, which is essentially a code of conduct for physicians that is still administered to new members of the profession at every medical school graduation.

The central advance we associate with the Hippocratic tradition is the notion that diseases have natural causes: they are not the result of the whim of the gods. This had the effect of separating the practice of medicine from religion. In fact, Hippocrates taught that the human body contained four "humors"—blood, black bile, phlegm, and yellow bile—and that illness was the result of these humors getting out of balance. This notion was to persist in medicine for over 1500 years, and survives today in our language, where we can refer to people being "bilious," "sanguine," or "phlegmatic."

The central strategy of Hippocratic treatment regimens was what we would call today "watchful waiting." He taught that the human body has enormous powers of recuperation, and that the physician's task is to facilitate those powers. In Hippocratic medicine, the concept of "crisis"—the point at which the body either starts to improve or moves toward death—played a major role. The teachings of his school were assembled in a series of texts (probably written by his students) in what is called the Hippocratic Corpus. These texts became major resources for later Islamic and European physicians.

It is clear that Greek physicians like Hippocrates benefited from the knowledge that had been created by their Egyptian predecessors. In China, the situation was quite different. There a unique brand of medicine developed, neither influenced by nor influencing physicians in the Mediterranean basin. As was the case with the Egyptians, we know about early Chinese medicine because of the survival of an ancient manuscript. In this case, the manuscript was titled *The Inner Canon of the Yellow Emperor*.

A word of explanation: the Yellow Emperor is the mythological ancestor of the Chinese (Han) people. He may have actually been a tribal leader in the Yellow River basin of China (a western analogy may be King Arthur of England, who may have been a Celtic tribal leader). Many inventions—all undoubtedly developed by other people—have been attributed to the Yellow Emperor, and the book about medicine surely falls into the same category. Like the Edwin Smith papyrus, it is a later copy of much earlier work—in the case of the *Inner Canon*, the original text was probably written sometime in the third millennium BCE, with existing manuscripts dating from 500–200 BCE.

As was the case with Hippocrates and his followers, Chinese medical theorists imagined the body as being controlled by a small number of elements. The elements that made up the Chinese universe as well as the human body were wood, fire, earth, metal and water. Different organs in the body were associated with different elements—the liver and gall bladder, for example, were associated with wood. Life energy, known as Qi (pronounced "chee" in Chinese, "kee" in Korean) flowed through the body in well defined channels.

Disease, in this picture, resulted from an imbalance of the body's components, and it was the physician's duty to determine, by examining the patient, what that imbalance was and how harmony could be restored.

As you might expect of a medical tradition that has been around for millennia, there is a large literature concerned with the application of these general principles. One example is the so-called "Eight Guiding Principles," which walk the physician through a set of opposite qualities to help with the diagnosis. (The qualities are hot/cold, deficiency/excess, interior/exterior, and yin/yang.) Many of the practices that we associate with Chinese medicine today—acupuncture, for example, or the extensive use of herbal remedies—have been part of this tradition from the beginning.

Islamic medicine

Medicine during the Islamic Golden Age grew out of the Greek and Egyptian traditions outlined above, with texts from the conquered Persian Empire playing a role as well. As was the case for science in general, Islamic physicians investigated many fields and made many advances over the centuries. In terms of lasting impact, I would like to point to three major innovations.

1 Hospitals in the modern sense made their first appearance during this period

There had always been places where people went for medical treatment, but in both Egypt and Greece such places tended to be places of religious worship. The great temple at Kom Ombo on the Nile, for example, was the "Mayo Clinic" for treating infertility in ancient Egypt. In the major Islamic cities you find, for the first time, large institutions for the treatment of the sick that were staffed by physicians who had well defined training regimens and ethical codes, but were not priests. Scholars have argued that the first hospitals in Europe (in Paris) were set up by leaders who had seen Islamic hospitals at first hand during the Crusades.

2 Islamic physicians established the communicability of some diseases

As we have said above, the basic theory of Islamic medicine derived from Greek ideas, and included the notion of humors we associate with Hippocrates. In the Greek tradition, disease resulted from an imbalance of the humors—it was, in other words, something that was internal to the body. By extensive study of what we would today call epidemiology, Islamic physicians established the fact that in some cases, disease could be transmitted from one person to another—that the cause of disease could be external to the body. We can recognize here an early precursor of the germ theory of disease that was rigorously established by Louis Pasteur in the nineteenth century (see Chapter 8) and is the basis of much of modern medicine.

3 Islamic physicians were skilled surgeons

Surgery was always a tricky business in a world without antibiotics and anesthetics, but Islamic surgeons developed many techniques to a high level. They routinely removed cataracts, for example, and had even developed reasonably successful procedures for more invasive processes like the removal of bladder stones.

Medicine flourished during the Islamic Golden Age for many reasons. One reason, of course, was just the opulent wealth of the Empire, which allowed courts to compete with each other for housing the most famous scholars. Indeed, visitors to modern day Istanbul are often struck, when visiting the tombs of fifteenth- and sixteenth-century rulers, that alongside of the list of battles the man had won is a list of famous scholars, poets, and physicians that were supported in his court. These sorts of things mattered in the Islamic world.

A second reason for this flowering, already alluded to above, is early Islamic writings. These writings guaranteed that there were no doctrinal roadblocks thrown in the way of medical advances. This was particularly important because it allowed Islamic physicians to perform autopsies and dissections, procedures forbidden in both Greek and medieval European societies. We will look at the lives and works of two of the most important Islamic physicians, Muhammad ibn Zakariya al-Razi (865–925), known as Rhazes in the west, and Abu Ali al-Husayn ibn Abd Allah ibn Sina (980–1037), known as Avicenna.

Rhazes was Persian, born in a town near Teheran. He had an unusual career path in that he probably began life as a professional musician (a lute player, according to one source), then took up the study of alchemy. He contracted an eye disease from contact with his chemicals, and his search for a cure brought him into the field of medicine. He became widely known in the Islamic world, and eventually became chief physician at the hospital in Baghdad. The way he chose a site for that hospital illustrates the empirical, data driven aspect of all Islamic science. According to legend, he went around to various potential sites in the city, hanging a piece of fresh meat at each location. He then chose the spot where the decay of the meat was slowest for his hospital. He was a pioneer in the study of infectious diseases, and it was at his hospital that he first differentiated between smallpox and measles, writing detailed descriptions of the progression of both diseases. He also wrote the first description of what we call hay fever today, in a charming chapter titled *Article on the Reason Why Abu Zayd Balkhi Suffers from Rhinitis When Smelling Roses in Spring*. He is also credited with being the first person to isolate ethanol, which you can think of as the invention of rubbing alcohol. There is actually some controversy about the etymology of the word "alcohol." The "al" prefix, of course, tells us that this is an Arabic word. In some dictionaries, the word is said to derive from "al-kuhul," where "kuhul" is the word for the substance we call kohl, which was used as an eye-liner in antiquity. The theory is that kohl was prepared by heating minerals in a closed vessel, and the word was extended to a substance derived by distillation. Modern scholars, however, point out that the Koran uses the word "al-ghawl," referring to a sprit or demon (think of the English word "ghoul"). The theory here is that the word refers to the ingredient of wine that produces intoxication. The confusion between these two similar words (al-kuhul and al-ghawl) by medieval translators would certainly be easy to understand.

Avicenna was also Persian, having been born in a province in what is now Uzbekistan. Contemporary accounts describe him as something of a prodigy, memorizing the Koran by the age of 10, becoming a practicing physician by the age of 18. He is supposed to have learned the Arabic numerals from an Indian greengrocer in his neighborhood. His adult life was complex—he was tossed hither and yon by various wars and dynastic struggles that roiled that part of the world. It's interesting that today the only reason that anyone remembers those wars is because they caused Avicenna to move from one court to another.

In any case, he eventually settled in Isfahan, in modern day Iran, dying there at the age of 56. Undoubtedly his greatest contribution to science was *The Canons of Medicine*, one of the most influential medical textbooks ever published. Just to get a sense of the book's longevity, we can note that it was still being used in European universities in the seventeenth century, more than half a millennium after it was written. It is a huge compendium of medical knowledge, 14 volumes in all, blending theory and practice, and it's not hard to understand the impact it had in medieval Europe when it was translated into Latin in 1472. It's hard to know where to begin in describing a voluminous work like this. The medical theory at the base of Avicenna's work is the Hippocratic notion of the four humors. To this was added the categorization of four "temperaments" (cold/hot, moist/dry) to make up a comprehensive classification scheme for patients. The anatomical descriptions and treatments in the *Canon* covered many fields of medicine. In the area of infectious diseases, for example, he argued that tuberculosis was transmitted from person to person, even though, following the Greeks, he identified the agent of infection as bad air. He is the first to have suggested quarantine as a way of controlling the spread of disease. He also identified some cancers as tumors, based on dissection. Contrary to many ancient traditions, he realized that the pulse was related to the heart and arteries, and did not vary from one organ of the body to the other. It was Avicenna who introduced the technique of taking the pulse from a patient's wrist, something we all still experience when we visit our doctors. He was also the first to give a detailed anatomical description of the human eye and to describe neurological diseases like epilepsy and psychiatric conditions like dementia. There is a story that he had as a patient a Persian prince who thought he was a cow and refused to eat. Avicenna is supposed to have pretended to be a butcher, telling the prince that he couldn't be slaughtered until he was fattened up. The prince began to eat again, at which point, the story goes, he recovered fully. Avicenna also thought a lot about what we would call today controlled medical trials. Undoubtedly aided by his training in logic and theology, he laid out a set of rules for testing new medications that would make a lot of sense to a modern practitioner. He stressed, for example, that drugs being tested should not be contaminated (he would have said "be free from accidental qualities"), and that (again in modern language) a large enough sample of data must be collected to allow statistically significant results (he would have cautioned against an "accidental effect").

Finally, the *Canon* contains a lot of advice that you could find in any health magazine today. He extols the value of regular exercise and proper diet, for example, and prescribes what we would call stretching before vigorous exercise. (One amusing sidelight: he included "camel riding" in his list of vigorous activities. You have to wonder how that was interpreted in medieval Europe.)

I could go on, but I think this should give the reader a pretty good idea of why the *Canon* played such an important role in the development of medicine in later ages.

Islamic astronomy

In mathematics and medicine, we have seen that cultures of the Islamic Golden Age played two important roles in the development of science. First, they preserved the works of classical antiquity, making them available to later scholars, and second, they made important advances on their own. My own sense is that in these two fields, the second function outweighed the first in importance. When we turn to astronomy, however, the situation is somewhat different. For reasons mentioned above, observations of the heavens played an important role in establishing the Islamic calendar, a fact that encouraged the growth of astronomy. Like the Babylonians and Greeks, Islamic astronomers compiled a large body of observational data. They did not, however, produce much in the way of astronomical theory, and they largely confined their work to elaborations and commentaries of the work of Claudius Ptolemy (see Chapter 3). Thus, in this case, the main contribution of the Islamic world to the advancement of science may have been, in fact, the transmission of Greek science to renaissance Europe.

In keeping with the tradition of polymaths in the Islamic world, many of the men we have already met made contributions to astronomy in addition to the work we've described. Our old friend al-Khwarizmi, for example, published an extensive catalogue of planetary positions in 830. Avicenna claimed to have observed a transit of Venus across the face of the sun in 1032 (although this claim is disputed by some modern scholars) and published a commentary on the *Almagest* in which he argued that stars were luminous, and did not shine by light reflected from the sun.

Although there were many astronomical observatories in the Islamic world, the observations tended to be confined to those needed for making the calendar and casting horoscopes. The most striking example of this restricted view of the heavens is the fact that the great supernova of 1054 (the one that produced the Crab Nebula) is not recorded anywhere in Islamic texts. The astronomers must have seen the supernova—the Chinese certainly did—and we can only conclude that they didn't think it was important enough to enter into their records.

Ptolemy had published a catalogue of about 1,000 stars and Islamic star catalogues rarely went beyond that list, although they did improve on Ptolemy's determinations of brightness (what we call magnitude). The first major revision of Ptolemy's star catalogue was by Persian astronomer 'Abd al-Rahman al-Sufi (903–86), who published his "Book of Fixed Stars" in 964. In this important text, the stars were given Arabic names, many simply translations from Ptolemy's original Greek, and it has usually been assumed that the names of stars like Altair, Aldebaran, and Rigel entered the west through al-Sufi's catalogue. Harvard historian Owen Gingerich, however, has argued that the names actually entered western astronomy through translations of Arabic books written as instruction manuals for the use of an instrument called an astrolabe. It's an interesting historical problem, but whatever the mode of transmission, the names survive to the present.

The greatest effort of Islamic astronomers was expended in refining and extending the work of Ptolemy, and here some problems peculiar to astronomy become important. Ptolemy had adjusted his complex wheels-within-wheels system so that it did a pretty good job of explaining the data that was available to him in his own time. Given the number of adjustable parameters at his disposal, it isn't surprising that he was able to do so. By the time Golden Age astronomers came on the scene almost a millennium later, though, the system was starting to show its age and things were beginning to get a little out of whack. One important task for Islamic astronomers, then, was to re-adjust Ptolemy's parameters to make his system work better, a task which they carried out very well indeed.

A more fundamental type of objection was raised by a number of leading philosophers working in Islamic Spain, including the Jewish scholar Maimonides (1135–1204) and the Muslim philosopher Abu 'l-Walid Muhammad bin Ahmad Rushd (known in the west as Averroes—1126–98). The problem had to do with one of the complications of the Ptolmaic system we've already mentioned, the fact that although the Earth was situated at the center of the universe, the crystal spheres didn't revolve uniformly around the Earth, but around a point away from the Earth called the "equant." In a perhaps over-simplified summary, we can say that these philosophers argued that if the Earth was to be the center of the universe, it ought to be the center of the celestial spheres as well. They urged their students to try to find a way to fix this problem. They couldn't, of course—they had too many facts stacked against them—but several of them gave it a good try.

In fact, sometime between 1025 and 1028, the Persian scholar Abu 'Ali al-Hazan ibn al-Haytham (known in the west as Alhazen) published a book whose title, *Doubts Concerning Ptolemy* pretty much tells the story. To him, the idea of the equant was simply unnatural. Here's how he makes that point in his book:

> Ptolemy assumed an arrangement that cannot exist, and the fact that this arrangement produces in his imagination the motions that belong to the planets does not free him from the error he committed in his assumed arrangement, for the existing motions of the planets cannot be the result of an arrangement that is impossible to exist ...

The Persian polymath Nasir al-Din al-Tusi (1201–74) was born in northern Iran and in 1259 supervised the building of a major observatory and astronomical center in what is now Azerbaijan. He developed an extension of the Ptolmaic model that, in essence, introduced extra epicycles to get rid of the equant problem (the device he used is called a "Tusi-couple"). Al-Tusi was the last of the great Islamic astronomers, and his followers used his system to bring the Ptolmaic model to as good a status as it could possibly achieve.

End of the Golden Age

The rapid decline of Islamic civilization and science followed a period of invasions by Crusaders from the West and, later, Mongols and Turks from the East. The question of why the decline was so precipitous remains a subject of debate among historians. The invasions were followed by centuries of petty civil warfare in various parts of the Islamic regions, and the resulting economic and social chaos surely played a role in the decline. In the end, though, we will have to agree with historian George Sarton when he said in a lecture titled *The Incubation of Western Culture in the Middle East* in 1950:

> The achievements of the Arabic speaking peoples between the ninth and twelfth centuries are so great as to baffle our understanding. The decadence of Islam and of Arabic is almost as puzzling in its speed and completeness as their phenomenal rise. Scholars will forever try to explain it as they try to explain the decadence and fall of Rome. Such questions are exceedingly complex and it is impossible to answer them in a simple way.

Transmission to the west

As intimated above, the Golden Age of Islamic scholarship came to an end in the thirteenth century, but not before much of the work we've been discussing had been translated into Latin. The center for this transmission was in modern Spain, where Muslim and Christian states shared a long (and shifting) border. Beginning in the twelfth century, the town of Toledo became the center for translation of works from Greek and Arabic, triggering what some scholars have called the "twelfth-century renaissance." At this time the town had recently been taken by the King of Castile, but still functioned as a multi-cultural center. It was, in other words, an ideal place for the exchange of knowledge to occur.

A major figure in the translation effort was a man named Gerard of Cremona (1114–87). Born in Italy, he traveled to Toledo, learned Arabic, and produced what was to become the most widely used and influential translation of the *Almagest*. Among the many other works he translated were al-Khwarizmri's *Algebra* and several of the medical works of Razes discussed above. Thus, just as Muslim scholars working in places like the House of Wisdom preserved the works of classical antiquity, Christian scholars in places like Toledo preserved the works of the Islamic Golden Age, bringing them to Europe where the next chapter in the development of science was to take place.

Chapter 6

The birth of modern science

Up to this point we have traced the development of various parts of the scientific method in many civilizations. We can think of this development as being a stream that flowed in countries around the Mediterranean and shifted to the Middle East during the Islamic Golden Age, with parallel but largely unrelated centers developing in places like China and the Americas. In this chapter we will trace the movement of that stream to a new location in northern Europe, which we have largely ignored up to this point. I will argue that from the fifteenth to the seventeenth centuries the scientific method in its fully developed modern form came on the scene in Europe, with the English scientist Isaac Newton being a good candidate for the title "First Modern Scientist."

You will recall that in Chapter 1 we outlined the scientific method as consisting of a never ending cycle of observation, theory, and prediction. Between the fifteenth and seventeenth centuries, a group of European scholars carried this cycle all the way around for the first time. The groundwork for this advance, of course, was laid down by the events we have been describing. The key point in the new advance was the growing realization that the theoretical models were taken to be exemplars of the real world, rather than just intellectual exercises. This meant that discrepancies between the theories and observations began to be taken more seriously. The days when a scholar like Maimonides could insist on having the planets move in circles, as we saw in Chapter 5, were over and a new age of empirical inquiry had begun.

After this development in Europe, the inherent international nature of the scientific enterprise asserted itself, and we will trace the beginnings of the new global science by examining the spread of the new ideas to the periphery of Europe, to Russia in the east and the American colonies in the west. In this discussion, we will concentrate on the developments of physics and astronomy, but we should keep in mind that advances were being made in many other fields as well.

The development of these sciences is a story most easily told through the lives of some of the individuals involved. In what follows we will look at the following figures:

Nicolaus Copernicus (Poland)
Galileo Galilei (Italy)
Tycho Brahe (Denmark) and Johannes Kepler (Germany)
Isaac Newton (England)

Nicolaus Copernicus (1473–1543)

Nicolas Copernicus was born into a prominent family in Poland—his uncle was a bishop with extensive connections in government and clerical circles. Marked early for a career in the Church, young Copernicus was sent off to study at the University of Bologna in Italy, where he completed degrees in both canon law and medicine. On his return to Poland, he was eventually appointed as canon of a large cathedral in the town of Fromborg.

A word of explanation: during the course of the Middle Ages wealthy and powerful people would often leave property to the Catholic Church on their death. As a result, cathedrals would accumulate a variety of possessions— land, farms, buildings, etc.—and the cathedral canon was basically the business manager for the enterprise. We know that Copernicus served on a commission to reform the Polish currency, and some scholars argue that he actually discovered Gresham's Law ("Bad money drives out good") before Gresham did. He also operated what amounted to a medical clinic at his cathedral, and may have led troops in various skirmishes with the Teutonic knights. Thus, Copernicus was very much a man of affairs in his native land, perhaps analogous to the mayor of a large city or governor of a small state in modern America.

It was customary during this period in late medieval Europe for prominent men to dabble in scholarly pursuits, more or less as a hobby. Translating obscure Greek poetry into Latin was popular, for example. For reasons we will never know, Copernicus chose astronomy as his discipline of choice. Although he built a small observatory in a corner of the cathedral yard, his main interest was not in making new observations. Instead, he set himself the task of answering a simple question: was it possible to construct a model of the solar system as accurate as Ptolemy's with the sun, rather than the Earth, at the center? Again, we have no understanding of how this idea came to him, so that issue will remain as another of those amiable but unsolvable historical mysteries.

He worked on his model for decades, sharing his thoughts with fellow European astronomers. Eventually, his friends prevailed on him to commit his theory to writing, and the result was the book *De Revolutionibus Orbium Coelestium* (On the Revolution of the Celestial Spheres). He sent the manuscript to Germany with a colleague to be printed, and according to some versions of the story, was actually handed a printed copy on his deathbed. But regardless of whether Copernicus ever actually saw the final version of his work, for the first time in 1500 years there was a serious scientific competitor to Ptolemy.

It is important to keep one thing in mind when we discuss *Revolutionibus*. Copernicus did *not* produce anything like our modern picture of the solar system. He was a transitional figure—a man with one foot in the future and the other firmly planted in the Middle Ages. He was willing to challenge one of the great unspoken assumptions of ancient astronomy—the idea of geocentrism—but unwilling to challenge the other—the idea of circular motion (see Chapter 4). Thus, when we look at the Copernican solar system, we don't see anything like our own. We do, indeed, have the sun at the center, but when we look at the planets, we see our old friends the epicycles and the equant back in play. In the end, despite his claims to the contrary, the Copernican system wasn't significantly simpler than Ptolemy's—it was just different.

One question that often arises in discussions of *Revolutionibus* involves the fact that the book did not cause any kind of disturbance with Catholic authorities when it was published. In fact, there were even seminars on the work given at the Vatican. Given the much harsher response to Galileo's popularization of the work, discussed below, this lack of response seems puzzling.

One important reason for this difference is that Copernicus, like all scholars of his time, wrote in Latin, and this had the effect of limiting his audience to the educated elite. In addition, Copernicus was a savvy Church politician, and, unlike Galileo, knew how to do things without ruffling feathers. In the introduction to his book, which Copernicus probably didn't write himself and many never have actually seen but would surely have approved, the reader is invited to consider the heliocentric model as a mathematical exercise, and not necessarily a representation of the real world. Whether Copernicus really thought that this is what *Revolutionibus* was about or not, it clearly had the effect of making the work less threatening to those in authority.

With Copernicus, then, we have our first step away from the unquestioning acceptance of the science of the ancient world, in this case a step away from the Ptolmaic universe. The work, however, has a much deeper meaning than that. By giving up the notion that the Earth was the center of the universe, Copernicus was, in effect, moving humanity away from the center of creation. When we look back at him, we don't bother with counting epicycles or looking at the details of his universe. Instead, we see the man who put us on the road that led, eventually, to our current understanding of human beings as one species inhabiting one planet circling an ordinary star in a low rent section of an ordinary galaxy, one galaxy among billions in the universe. In fact, scientists today often speak of what is called the "Copernican Principle," which is the idea that there is really nothing special about the Earth or those of us that inhabit it—that we are all quite ordinary parts of an unimaginably large universe. Deep conclusions from the results of a busy man of affairs' hobby!

Tycho Brahe (1546–1601) and Johannes Kepler (1571–1630)

Like Copernicus, Tycho Brahe was born into a position of privilege, in Tycho's case, into a noble family in Denmark. His adventurous life began shortly after his birth. He was one of a set of twins, and his father had promised Tycho's uncle that the uncle would have one of the boys to raise. Tycho's brother died shortly after their birth, so his uncle walked off with Tycho when the boy was about two years old, obviously feeling that this was his share of the bargain.

According to his later writing, young Tycho became fascinated with astronomy because of a solar eclipse. It wasn't so much the majesty of the eclipse itself that attracted him, however, but the fact that someone had been able to predict exactly when it would occur. Despite the objections of his family, he undertook a study of astronomy during his student years. It was during those years that another unusual event happened. He fought a duel with a fellow student—over who was the better mathematician, according to one source—and lost the tip of his nose. For the rest of his life he wore prostheses made of metal.

In 1572 a rather extraordinary event happened in the skies; a new star appeared in the constellation Cassiopeia where none had been before. In the Ptolmaic system such an event was impossible, since the heavens were supposed to be eternal and unchanging. The only way out for traditional astronomers was to argue that the new star (what we would call a supernova today) was closer than the moon, in the sphere where change was allowed. Tycho realized that if the supernova were that close, it would appear to move against the background of the fixed stars as the moon did. When his highly accurate measurements showed that this didn't happen, Tycho became famous as one of Europe's leading astronomers.

The King of Denmark was so pleased that a member of his court had achieved such stature and acclaim that he gave him an island out in the sound between Denmark and Sweden and funds to build a major astronomical observatory. For over twenty years, Tycho worked at that place, called Uraniborg, and assembled the best data ever taken on the movements of the planets.

Tycho lived before the invention of the telescope, and was the last of the great naked eye astronomers. In those days astronomers measured the positions of the stars and planets by a technique analogous to sighting along a rifle barrel. What made Tycho's measurements so unusual was the fact that he was the first astronomer to think seriously about how his instruments introduced errors into his measurements. For example, as temperatures dropped during a night of observing, the various parts of his instruments would start to shrink, usually at different rates. Tycho took these kinds of effects seriously and worked out how to correct his readings. Over the years, then, as mentioned above, he put together the best collection of data ever taken on the movements of the planets.

Toward the end of his career, Tycho had some serious disagreements with the new King of Denmark, and as a result moved his operations to Prague. There he hired a young German mathematics teacher by the name of Johannes Kepler to help him analyze his data. By modern standards, Kepler was something of a mystic. He seems to have been obsessed with the notion that the universe was a work of art. His first attempt to produce a universe in keeping with this idea involved seeing the planets as a geometrical work of art, fitting the planetary orbits into a geometrical pattern. This didn't work out; the orbits just aren't that simple. Later in life, after he did the work described below, he got the idea that the universe was a musical work of art, and actually wrote on the subject of which planet sings soprano, which tenor, and so on. In between these two flights of fancy, however, he sat down with Tycho's compendium of measurements and worked out what the orbits actually are. I suppose that you can think of this as seeing the universe as an intellectual work of art.

With Tycho's massive database, Kepler was able to ask a simple question: what is the actual shape of a planetary orbit? Remember that up to this time everyone had simply assumed that the orbits had to be circular, and added epicycles to make their assumptions match the data. Kepler, in essence, dropped the unspoken assumption of circular motion in the heavens and let the data tell him the shape of the orbits, circular or not. To his (and everyone's) surprise, they turned out to be elliptical. In the end, he wound up enunciating three laws that described the solar system—statements now known as Kepler's Laws of Planetary Motion. In somewhat simplified form, they are:

The planets move in elliptical orbits
They move faster when they are near the sun than when they are farther
     away
The farther away from the sun a planet's orbit is, the slower the planet
     moves.

We don't have time to discuss these laws in detail, but the point is that with Kepler's work we have finally gotten rid of both of the unspoken assumptions of the Greeks. Although it would be some time before these ideas were fully accepted, between Copernicus and Kepler we have something that looks like our current view of the solar system. Having said this, however, we need to note that Kepler's Laws are purely empirical. We know what the planets do, but we don't know why they do it. For that next step, we turn to the work of two extraordinary men—Galileo Galilei and Isaac Newton.

Galileo Galilei (1564–1642)

Galileo was born in the Italian city of Pisa, then part of the Duchy of Florence. He studied mathematics, and in 1592 he became a professor of mathematics at the University of Pisa, where, during the next 18 years, he did his most important work. He has an unusual place in history, because, from the point of view of the development of science, he is best known for the wrong reason. Most people associate Galileo with his trial on suspicion of heresy near the end of this life. We'll get to that trial below, but before we do, we will look at two major scientific advances he made, advances that have earned him the title of "Father of Experimental Science." These two advances were (1) the beginnings of our modern ideas about the motion of objects, and (2) the first telescopic survey of the heavens. Let's take these in order.

If I tell you that a car is moving at the rate of 40 miles per hour and asked you where the car will be an hour from now, no one would have trouble saying "40 miles away." Handling uniform motion of this type is simple, and the problem could easily have been solved by any of the ancient mathematicians we've discussed. The situation changes, however, if we let the car accelerate. (The easiest way to think about this is to visualize a speedometer dial—for uniform motion, the dial doesn't move; for accelerated motion it does.)

The problem of accelerated motion had bedeviled mathematicians for centuries, and some approximate computational schemes had been devised, but Galileo was the first to understand it in a fundamental way. He knew that falling bodies pick up speed as they approach the ground, and he used this knowledge to set up an ingenious experiment. He would roll a large metal ball down an inclined plane on which were placed a series of tightly stretched wires. As the ball rolled down, there would be a series of "pings" as it crossed the wires. Galileo could adjust the positions of the wires until the time intervals were the same (the human ear is very good at judging equal time intervals). With these experiments, he established that the mass of the ball didn't matter—heavy ones went down at the same rate as light ones—and that the velocity increased linearly as the balls rolled down; a ball would be moving twice as fast after two seconds as it was after one, for example. From this, he was able to show that the path of a projectile thrown upward would be a parabola. With Galileo, then, the basic laws of accelerated motion were laid out.

It's important to realize that these results completely overturned the Aristotelian science that had been the mainstay of intellectual study for generations. Aristotle had taught that heavier bodies would fall faster than light ones, for example. It's also important to realize that Galileo never resolved the Aristotelian problem of natural and violent motion (see Chapter 4). He said, in essence, that he could tell you where a projectile was and how fast it was moving at any point on its trajectory, and that the question of natural and violent motion never entered the picture. The Aristotelian categories, in other words, were simply inappropriate in describing projectile motion.

Incidentally, there is no evidence that he dropped weights off the Leaning Tower of Pisa to prove his point. Had he done so, the effects of air resistance would have caused the heavier object to hit the ground before the lighter ones.

His studies of motion earned him high prestige, but it was his astronomical use of the telescope that vaulted him to international prominence. He didn't invent the telescope, but he built an improved model based on a description of earlier instruments. In late 1609 and early 1610, he was the first man to turn the telescope to the skies. His discoveries were amazing. He saw mountains on the moon, sunspots, and the four largest moons of Jupiter, among other things. All of these phenomena were in direct conflict with the prevailing Ptolmaic picture of the universe.

In the Ptolmaic system, the spheres away from the Earth were supposed to be pure and unchanging, yet here was evidence for "imperfections" like mountains and sunspots where none were supposed to be. The really damning evidence against the system was the moons of Jupiter, however. The basic premise of ancient philosophy was that the Earth was at the center of the cosmos, and everything else revolved around it. Yet here were four objects that seemed perfectly happy to orbit a center that wasn't the Earth. Galileo named the moons the "Medicean Stars," in honor of the Grand Duke of Tuscany, and was rewarded with a lifetime position at the Duke's court.

Galileo published his findings in a book titled *The Starry Messenger* in 1610, and this is where his troubles with the Church started. He wrote in Italian, which made his work accessible to people who didn't read Latin. His championing of the Copernican system triggered a response, and in 1616 he may or may not have been warned not to "hold or defend" the system, although he was allowed to study it as a mathematical exercise. (The historical evidence on what actually happened in 1616 is ambiguous.) In 1632, he published his *Dialogue Concerning the Two Chief World Systems*, putting forward what turned out to be an incorrect theory of the tides as evidence for the Copernican system. He also put the arguments of his onetime friend the Pope into the mouth of a character named "Simplicio" (which translates roughly as "Fool") – small wonder that his many enemies had such an easy time convincing the Church to proceed against him. In what has become an iconic moment in history, he was tried on suspicion of heresy and allowed to purge the suspicion by recanting his views. He spent the rest of his life under house arrest in his villa near Florence, continuing his experimental work.

Isaac Newton (1642–1727)

It's a little difficult to characterize the social position of Isaac Newton's family. They owned a farm and worked part of it themselves, but also had tenant farmers. This would put them somewhere between a Midwestern family farm and an agribusiness operation in modern America. In any case, Newton went off to Cambridge University as a young man.

As it happened, 1665 was the year when the Black Plague made its last appearance in England. People fleeing infected cities spread the disease, and the situation became so serious that the authorities at Cambridge closed the university for 18 months to let the plague burn itself out. Newton went back to his family home, and those 18 months became one of the most productive periods in the history of science. Thinking alone, he derived (1) the differential calculus, (2) the integral calculus, (3) the theory of color, (4) the laws of motion, (5) the law of universal gravitation, and (6) proved several miscellaneous mathematical theorems. It's no wonder that scientists agreed with author Isaac Asimov when he said "When scientists argue about who was the greatest scientist, they are really arguing about who was the second greatest."

For our purposes, we need consider only the laws of motion and the law of universal gravitation. Deceptively simple, the laws of motion describe any motion of any object anywhere in the universe. They are:

I   An object in motion will move in a straight line unless acted on by a force

II   The acceleration of an object is directly proportional to the force and inversely proportional to the object's mass

III For every action there is an equal and opposite reaction.

The first of these laws overturns a millennium of thinking about motion. Had you asked Ptolemy why the heavenly spheres kept turning, he would have said that an object left on itself will move in the most perfect geometrical path a circle. The First Law says otherwise and, in effect, replaces the circle with a straight line.

There is, in fact, a simple thought experiment you can do to see what Newton was driving at. Imagine tying a rock to a string and twirling it around your head. So long as you hold on to the string, the rock moves in a circle, but if you let go it moves off in a straight line. The Newtonian interpretation of this experiment is simple: the rock wants to move in a straight line, but the force exerted by the string keeps pulling it back into a circle. You can actually feel this force in your hand. Letting go removes the force, so the rock is free to move in a straight line.

The First Law, then, tells you when a force is acting. If you see a change in the state of motion of an object—and acceleration, deceleration, or change of direction, then you know that this is the result of a force. The Second Law tells you in more detail what the force does. The bigger the force, the higher the acceleration: the heavier the object is, the harder it is to change its state of motion. It's easier to deflect a ping pong ball than a bowling ball, for example.

The Laws of Motion, then, tell you about how objects react to forces, but don't talk about any force in particular. To talk about a specific force, we need to look at the law of universal gravitation.

Later in his life, Newton described the events that led to the law this way: he was walking in the apple orchard one day when he saw an apple fall from a tree, and behind the tree he saw the moon in the sky. He knew that the apple fell because of a force that was called gravity, but he also knew (remember the rock and the string) that there had to be a force acting on the moon. If there weren't, the moon would fly off the way the rock does when you let go of the string. At that moment, he asked the kind of question that seems simple in retrospect, but requires real genius to ask for the first time: could the force that made the apple fall be the same force that kept the moon in orbit?

As it happens, they are the same force. Instead of the old Aristotelian system in which there was one force making things fall on Earth and another operating out among the crystals spheres, we have a single force operating everywhere. Newton's discovery unified the heavens and the Earth, and it was the first of many such unifications we'll discuss in later chapters. For the record, the Law of Universal Gravitation can be stated as follows:

> Between any two objects there is an attractive force that is proportional to the product of their masses and inversely proportional to the square of the distance between them.

Using this law and combining it with the laws of motion, Newton was able to work out the orbits of the planets, deriving Kepler's empirical laws in the process. Thus, Kepler's Laws became a consequence of universal gravitation, rather than merely an empirical set of rules. Newton was also able to help Halley work out the orbits of his comets, as discussed in Chapter 1. It is for this reason that I have proposed that we look at him as the first modern scientist.

Newton's influence was not confined to science, anymore than Copernicus' was. The Newtonian universe was a wonderful, orderly place. The planets moved around like the hands on a clock, with Newton's Laws and calculations defining the "gears" that made the whole thing go. In the language of the seventeenth century, the human mind could, by the use of reason, find out what was in the mind of God when He created the universe.

You can see the influence of this idea of an orderly, mechanical universe in the literature and music of the time, and it certainly played a major role in bringing about the Enlightenment. Some scholars have even argued that it had an influence on the American constitution. The idea is that the Founding Fathers had been educated into the Newtonian system and said, in essence, that "Newton discovered the laws that make an orderly universe, and we will discover the laws that make an orderly society." We will see this sort of effect, in which ideas from the sciences spread out into the larger culture, many times in subsequent chapters.

With Newton, then, we have finally arrived at the modern scientific method. Before we go on to see how this method of inquiry progressed, I would like to consider two things: First, I would like to discuss the way that the method moved away from its original home in northern Europe and began to spread around the world. Before that, however, I would like to ask what historians have come to call the "Needham Question." In our context, the Needham Question can be stated as "Why did modern science develop in northern Europe and not somewhere else?"

The Needham Question

Joseph Needham (1900–1995) began his academic career as a biochemist at Cambridge University in England. During the 1930s, his interactions with visiting Chinese scientists triggered a deep interest in Chinese culture—he is supposed to have learned the language from one of those visitors, in fact. During World War II he served as director of a Sino-British science office in China, and stayed on to study the history of China, particularly the history of Chinese science and technology. He began editing a massive series of books under the title of *Science and Civilization in China*, a series that was continued by Cambridge University Press after his death.

As a historian, Needham could scarcely miss the fact that Chinese science had been the equal of science in the Mediterranean and Islamic worlds, yet it was quickly surpassed by the events in Europe we've been describing. The question of why this happened has come to be known as the "Needham Question" (sometimes stated as "Needham's Grand Question"). Needham's favorite answer to his question had to do with what he saw as the stifling effect of Confucianism and Taoism in China. His argument was that the excessive traditionalism inhibited the kind of individual initiative that developed in Europe. To take just one example of this effect, he described a tradition in China that students who question their teachers are seen as showing disrespect to their elders. Since science depends on a constant questioning, the argument goes, students brought up in this way are unlikely to initiate important new research.

Other answers to the Needham Question have been offered. A popular one involves the fact that the Chinese language does not have an alphabet. The suggested effects of this situation range from the difficulty it imposes on establishing a printing industry to hypothetical effects of the existence of an alphabet on higher cognitive functions in the brain. (Given the successes of modern Chinese and Japanese scientists we'll discuss later, I think we can safely discount this last suggestion.)

A more generalized form of the Needham Question would be something like "Why did modern science develop in Northern Europe and not somewhere else?" People who ask this question usually have either China or Islamic societies in mind as alternate venues. Those who ask the Needham Question about Islam often cite the tendency of Islamic religious leaders to blame societal reverses on a falling away from an earlier, purer form of their religion.

This tends to lead to various kinds of theocracy, the argument goes, in which little value is placed on new knowledge of the type developed by science.

As to why modern science developed in Europe, there are again many kinds of speculative answers advanced. One, already alluded to in Chapter 5, involves the rise of Protestantism in Europe, leading to a questioning of accepted forms of authority. The Protestant emphasis on the relationship of an individual to God, without the necessity of the intervention of an established ecclesiastical authority, is seen by many as constituting an enablement of the individual achievement and responsibility that is the cornerstone of modern science.

Another frequently mentioned factor involves European geography, which encourages the creation of smallish political entities, each protected by natural boundaries like rivers and mountains. The idea here is that in a continent split up into small (and frequently hostile) political entities, it is impossible to establish the kind of suffocating centralized orthodoxy that Needham saw in China and earlier observers saw in the Ottoman Empire. The early history of modern science is full of examples of Catholic scholars moving to Catholic countries while their Protestant counterparts moved to Catholic countries.

Finally, the rise of Humanism that accompanied the re-discovery of classical knowledge in the Renaissance is often mentioned as a precursor to the scientific revolution. By questioning the established Scholastic orthodoxy, the Humanists cleared the way for later scholars who, ironically, ended up destroying the science of Aristotle and the other giants of antiquity.

In any of these kinds of historical situations, it is perfectly reasonable for historians to seek out specific causes for specific events. As it is usually asked, however, the Needham Question is often an exercise in what is called "counterfactual history"—"What if X had happened instead of Y?" From my point of view, asking this sort of question about the development of science is rather pointless. The laws of nature are out there, and they are sure to be discovered by someone somewhere. If science hadn't developed in northern Europe in the seventeenth century, it would have developed somewhere else at a different time. Had that happened, we would then be asking the Needham Question about a different place and time ("Why Iceland?"). The situation reminds me of an old country and western song, which said "Everybody's got to be somewhere, so I might as well be here."

Scientific societies

The term "scientist" was an invention of the nineteenth century. In the seventeenth century, people who did what we think of as science were called "natural philosophers." Before the nineteenth century, only people of independent means had the luxury of spending their time thinking about science—a very different situation than the one we're used to today. This means that the natural philosophers of Newton's time were spread rather thinly on the ground.

Today we know that the process of scientific inquiry is very difficult to carry out in isolation. Modern scientists routinely communicate with their peers through journals and at numerous meetings. Scientists in the seventeenth century had the same need for communication, and therefore built institutions to supply that need. One way we can trace the dispersal of the scientific method from its original home is to discuss the spread of these institutions.

The oldest scientific organization is the Royal Society of London for the Improvement of Natural Knowledge, usually referred to as the Royal Society, which was founded by Royal Charter in 1662 by King Charles II. The society was a formal version of a loosely knit group of about a dozen scientists known as the "Invisible College" which had been meeting in various locations around London for some time. Initially, the meetings of the society were devoted to running demonstrations and experiments for the members, but eventually the publication of a journal for the communication of new results was undertaken.

Over the years many famous scientists have been members of the Royal Society—Isaac Newton himself served as President of the Society from 1703 until his death in 1727. Later, Charles Darwin (see Chapter 8) read his now famous paper on the theory of evolution to the same body. Today, the Royal Society is heavily involved in promoting scientific research and in advising the British government.

The success of the Royal Society quickly spurred similar developments elsewhere. In 1700, following the advice of the German scientist and mathematician Gottfreid Leibniz (1646–1716), the then Elector of Brandenburg chartered a similar society, which became the Royal Academy of Sciences when the Elector became King of Prussia a year later. It is interesting that he decided to fund the Academy by granting it the exclusive right to produce and sell calendars in his kingdom.

Incidentally, Newton and Leibniz are another illustration of the point we made in Chapter 5 that major discoveries are often made independently in different places by different people. Both of these men developed the calculus and, unfortunately, engaged in a bitter lifelong battle over precedence. The whole business looks a little silly to the modern eye—there was certainly enough credit to go around—but it triggered a great deal of nationalistic animosity at the time.

In any case, Leibniz went on to be an advisor to Tsar Peter the Great in Russia, and in 1724, after his death, was influential in founding the Saint Petersburg Academy of Sciences. Many of the great scientists and mathematicians of the nineteenth century visited or worked in what was then the capital of Russia. After the Russian Revolution, the organization was reconstituted as the USSR Academy of Sciences and, in 1934, moved to Moscow. After the collapse of the Soviet Union, it became once again the Russian Academy of Sciences.

While a scientific academy was being established at the far eastern edge of Europe, another was established in a European colony in the New World. In 1743 the American Philosophical Society was founded in Philadelphia under the guidance of Benjamin Franklin. Many of the great leaders of America were members—George Washington, Thomas Jefferson, and James Madison, for example—as well as many prominent foreign luminaries like Alexander Humboldt and Tadeusz Kosciuzsko. The Philosophical Society is still functioning in Philadelphia—indeed, the author has had the privilege of speaking there.

Thus, once modern science started in the heart of Europe, it quickly spread to the European periphery. In subsequent chapters, we will see it spreading around the world to become a truly global enterprise.

Chapter 7

The Newtonian world

With the development of the modern scientific method, two important goals had been achieved: first, scientists had learned to ask the right kind of questions about nature and, second, they had hit upon a systematic way of attacking and answering those questions. To take the familiar example of projectile motion, it was now recognized that the old issue of natural vs. violent motion arose as a result of human beings setting up categories that had nothing to do with nature. The problem arose because of the way we look at the world and had nothing to do with nature itself. Once the question was rephrased—where will the projectile actually go, for example—progress could be made.

As a result of these advances, the eighteenth and nineteenth centuries saw an explosion of scientific knowledge, centered in the home grounds of western Europe. In this chapter we will look at a few of these advances. We will, of course, see the methods discussed in the last chapter applied to areas of science other than mechanics—most notably to the fields of chemistry and electricity and magnetism. We shall also see a new phenomenon—the impact that developments in science can have in transforming society in fundamental ways. The electrical generator is a familiar example of this phenomenon, but there are others—think of the telegraph, for example. Similar advances in the biological sciences will be dealt with in the next chapter.

Alchemy, chemistry, and the new theory of the atom

Alchemy is a term that comes to us from the Arabic *al-kimia*, with many scholars arguing that *kimia* comes from an old Egyptian word for their country (it apparently referred to the "dark earth" along the Nile, as distinct from light colored dirt of the desert.) Every culture we've talked about practiced alchemy in one form or another, and it is generally thought of as the precursor to the modern science of chemistry.

The title of a text by the German alchemist Paracelsus (1493–1541) pretty much describes what alchemy was about. The book was called *Solve et Coagula* (take apart and put together), and this is what alchemists did with materials at their disposal. Whether we're talking about an Egyptian priest preparing ointments for mummification, a Babylonian mining engineer figuring out how to get metal from ore, or a Chinese physician mixing herbs to create a medicine, the process is the same. Materials are mixed together and treated in some way—by heating, for example—and something new emerges. Over the millennia, an enormous amount of empirical knowledge about what we would call chemical reactions was accumulated in this way.

There are generally reckoned to have been three great goals for alchemists. The first (and the one most people have heard about) was the search for the "philosopher's stone," which could change lead into gold. The second was the "elixir of life" which could cure any disease and make the user immortal. The third, more technical in nature, was the search for a universal solvent, capable of dissolving any material. Needless to say, none of these searches was successful, as a moment's reflection will convince you they had to be. I mean, if you found the universal solvent, what would you put it in? Nevertheless, by the eighteenth century there was a lot of empirical knowledge available.

The man who is usually associated with moving chemistry from a collection of empirical facts to a modern science is the Frenchman Antoine Lavoisier (1743–94). He was born into a wealthy family in Paris and well educated. A brilliant man, he was elected into the French Academy at age 25 (because of a paper he wrote on street lighting, of all things!). It was Lavoisier, more than anyone, who took the vast, chaotic, unorganized knowledge of chemical reactions that were the legacy of the alchemists and other predecessors and incorporated them into a single coherent theory. His book *Elementary Treatise on Chemistry* (1789) became the Bible of the new science. For our purposes, we will look at two of Lavoisier's achievements—the law of conservation of mass and the notion of the chemical element.

The conservation of mass (in modern terms) says that the total amount of mass (or, equivalently, the number of each type of atom) involved in a chemical reaction does not change from the time the reaction starts until it ends. Think back to your high school chemistry, when you had to balance chemical equations. You probably recall that if a reaction started with three oxygen atoms, it had to end with the same number. This was, in effect, an application of the principle that Lavoisier first introduced into science.

His discovery of the conservation of mass grew out of his insistence on making precision measurements of the amount of different materials involved in an interaction. In a classic experiment, for example, he sealed some fresh fruit slices in a jar and carefully weighed the jar and contents. A week later the fruit had turned into a disgusting goo on the bottom of the jar and there were drops of water on the inside of the glass. Despite all the changes that had obviously taken place, however, the weight of the jar was exactly the same as it had been on the first day. By careful measurements of this kind, Lavoisier established the fact that (again in modern language) even the most complex chemical processes involve a shuffling around of a fixed number of atoms, with nothing new being either created or destroyed. We'll encounter this concept again in Chapter 9, when we talk about the work of Albert Einstein.

Lavoisier did not so much discover the existence of chemical elements as incorporate them into his comprehensive theory. Part of the empirical knowledge that had been gained was that, while most materials could be broken down by chemical means (burning, for example), there were a few that could not. These fundamental substances were called elements. Lavoisier discovered a few (oxygen and hydrogen, for example), and there were some 20-odd others known at the time. The notion that the elements were somehow more fundamental than everything else became a cornerstone of the modern atomic theory.

In addition, unexplained regularities began to be seen. For example, if water was broken down into its elements, it always turned out that there was eight times as much oxygen (by weight) as hydrogen. This so-called "Law of Definite Proportions" held for many materials, but couldn't be explained.

Although Lavoisier's birth into a prominent family was obviously an advantage in his scientific career, it was not an advantage after the French Revolution. He had been a member of a prominent (and highly unpopular) private tax collecting company, and during the Reign of Terror in 1794 he was accused of treason and guillotined on May 8. An international appeal to spare his life was met by the judge with a statement that has become all too typical of revolutionaries of both the right and left since then: "The revolution needs neither scientists nor chemists." Perhaps the best statement about Lavoisier's execution was made by the Italian born physicist and mathematician Joseph Lagrange: "It took only an instant to cut off his head, but France may not produce another like it in a century."

As France fell into chaos, the story of atomic theory moved across the Channel to England. Let's take it up with John Dalton (1766–1844), a teacher in a Quaker school in northern England. It was a peculiarity of the English educational system at that time that no one could enter a university unless he subscribed to the Articles of Faith of the Anglican Church. As a Quaker, Dalton couldn't do that, so he was educated informally, outside of official channels. He was interested in a number of scientific questions, like color blindness (sometimes called Daltonism in his honor), but what is important to us was his studies of atmospheric gasses. Motivated by these studies and with things like the Law of Definite Proportions and the existence of chemical elements in mind, he formulated the first modern atomic theory.

Basically, Dalton realized that all of the great amorphous collection of chemical knowledge could be explained by a simple idea, what we will call the modern atomic theory to distinguish it from the philosophically driven system of the Greeks. The basic tenets of the theory are as follows:

Matter is made of indivisible particles called atoms
All atoms of a given chemical element are identical
Atoms of different chemical elements are different from each other

This simple theory explains a great deal of the accumulated chemical knowledge. Most materials are compounds, made by putting various collections of atoms together. For example, water is made from two hydrogen atoms and one oxygen atom. The oxygen weighs 16 times as much as hydrogen, so if you take the water apart you'll find 8 parts oxygen to 1 of hydrogen. Thus, the Law of Definite Proportions is readily explained as well. All of this was published in a book titled *New System of Chemical Philosophy* in 1808.

Dalton's model, based as it was on empirical evidence, was a huge step forward in understanding the basic structure of the universe. As the nineteenth century progressed, howver, a number of events happened that suggested that the real atomic theory was somewhat more complicated than Dalton had originally thought.

- *More elements were discovered:* When Dalton published his work people knew of some 26 elements. As the nineteenth century progressed, this number climbed quickly (today we're at 118 and counting). This development detracted from the original simplicity of the Dalton picture, but didn't disprove it.
- *Unexplained regularities were found:* The most important of these was the discovery of the periodic table of the elements by the Russian chemist Dimitri Mendeleyev (1834–1907) in 1869. While compiling data for a chemistry text, he found that the elements fit naturally into a scheme of rows and columns in which the weight increased from left to right in each row while elements in the same column had similar chemical properties. The system worked, but until the advent of quantum mechanics (see Chapter 9), no one knew why.
- *The atom turned out to be divisible after all:* In 1895, the British physicist J.J. Thompson identified the particle we now call the electron as a constituent of the atom (see Chapter 9). Far from being indivisible, then, the atom had an internal structure that would be the focus of physics in the twentieth century.

While this elaboration of the atomic model was going on, a curious debate started among European scientists. Given that matter seemed to behave *as if* it was made up of invisible atoms, did it follow that atoms are real? Couldn't they just as well be constructs of the human mind, with no actual physical existence? How are scientists supposed to deal with this kind of question?

The most prominent physicist involved in this debate was Ernst Mach (1838–1916). Born in what is now the Czech Republic, he eventually wound up as a professor at the Charles University in Prague, which was then part of the

Austro-Hungarian Empire. He worked in many fields—in fact, he once turned down the offer of a Professorship of Surgery—but is best known for two things: having the speed of sound in air named after him (an airplane traveling at this speed is said to be doing "Mach 1") and his opposition to the reality of atoms.

Mach was too good a scientist to deny that Dalton's picture gave an elegant and useful way of thinking about the world. His objections were more philosophical in nature. In fact, Mach was one of the founders of the field that is now called the "philosophy of science."

He was a strong proponent of a philosophical point of view known as positivism, which holds that the only meaningful questions in science are those that can be decided by experiment or observation. Since atoms couldn't be observed, Mach argued, they should not be regarded as real, although he acknowledged that they were a very useful concept. There is a long history of debate between Mach and prominent nineteenth-century physicists on this question.

Surprisingly, the debate was resolved in 1905 when a young clerk in the Swiss Patent Office named Albert Einstein published a paper on an obscure topic called Brownian Motion. The effect is named after British botanist Robert Brown (1773–1858) who published, in 1827, a description of what he saw in his microscope when he suspended small pollen grains in a drop of water. Instead of sitting still, the pollen grains jittered around in what looked like random motion.

What Einstein realized was that the pollen grains were constantly being bombarded by molecules of water. Each time a molecule bounced off, it exerted a small force on the grain. On average, these forces should cancel out—as many molecules should bounce to the right as bounce to the left, for example. At any given instant, however, ordinary statistical fluctuations will ensure that more molecules will hit on one side than on the other. In that instant, then, there will be a net force on the pollen grain, causing it to move. An instant later, the situation may be reversed and the force exerted by the molecules will cause the grain to move back. This, of course, was exactly the kind of jittery motion Brown had seen. Einstein worked out the mathematics and showed that the motion could be completely explained in this way.

Here's the point: mental constructs can't exert forces, only real material objects can. With Einstein's explanation of Brownian motion, then, the debate on atoms came to an end. They were real objects, capable of exerting forces. Period. With this development, the attention of scientists turned to understanding the structure of the atom, as we shall see in Chapter 9.

Electricity and magnetism

The existence of the basic phenomena associated with electricity and magnetism was known to most of the ancient cultures we'll talk about. They never constituted a major subject of scientific inquiry, however—think of them as sort of an intellectual side show compared to subjects like medicine and astronomy. It wasn't until the scientific revolution that a really serious study began, and some unexpected discoveries, outlined below, turned the field from an obscure corner of science into a major driver of modern technological society. In this section, we will trace the development of the knowledge of electricity to the eighteenth century, then go back and do the same for magnetism, and finally discuss the development of a single coherent theory that embraced both.

Electricity first. The Greeks knew that if you rubbed a piece of amber with cat's fur, it would attract small bits of material. (The word "electricity" comes from the Greek word for "amber.") You can get the same effect by running a comb through your hair on a dry day and letting it pick up little pieces of paper. We say that this is an effect of static electricity, and we say that the amber or comb acquires an "electrical charge" in the rubbing process.

More to the point, the Greeks knew that if you touched two pieces of cork to the amber, those pieces of cork would repel each other. Please note that this means that the new force we're dealing with is different from gravity. Gravity is *always* attractive—it can never be a repulsive force. This fact will be important in Chapter 11, when we talk about the fundamental forces of nature.

In addition, the Greeks knew that you could get very similar effects if, instead of rubbing amber with cat's fur, you rubbed a piece of glass with silk. Bits of cork touched to the glass would repel each other. On the other hand, if you took at bit of cork that had been touched to the amber and brought it near a piece that had been touched to the glass, they would be attracted to each other.

In modern language, we can summarize these results in the following way:

1 Rubbing materials together results in producing electrical charge.
2 There are two kinds of electrical charge, which we customarily refer to as "positive" and "negative."
3 Like charges repel each other, unlike charges attract each other.

These general precepts represent the state of knowledge about electricity from ancient times to the eighteenth century. By that time, of course, the modern scientific method had been enunciated, so investigators had a road map about how to proceed. Throughout the century, researchers developed ways to accumulate large static charges and investigated the way that charged bodies behaved. Amusingly enough, they called themselves "electricians," a word that has acquired quite another meaning in the modern world. We will look at two "electricians" in what follows—Benjamin Franklin (1706–90) in the American colonies and Charles Coulomb (1736–1806) in France.

Benjamin Franklin, of course, needs no introduction to American readers. His name calls up images of a bald, bespectacled man playing out an important political role in the founding of the new country. Less well known is that Franklin was a major figure in the study of electricity in the eighteenth century. It is no exaggeration to say that had Nobel Prizes been awarded back then, he would have received one. We will examine two of his contributions, one in theory and one in technology.

Franklin was the first to realize that the accumulated knowledge about static electricity could be explained in terms of the movement of one kind of charge. For pedagogical reasons, I will explain his so-called "single fluid theory" in modern terms that Franklin would surely not have recognized. We will see in Chapter 9 that materials are made from atoms in which negatively charged electrons circle in orbits around positively charged nuclei. Normally, there are as many negative as positive charges in the atoms in any material, so the material is electrically neutral. When we rub that material (remember the amber and cat's fur), one of two things can happen. First we can pull electrons out of the material, leaving behind an excess of positive charge. Alternatively, we can push electrons into the material, leaving it with an excess of negative charge. Thus, the motion of a single type of charge (which we call electrons and Franklin would have called an electrical fluid) can explain the existence of the two types of electricity. This realization set the stage for later developments, as we shall see.

On the practical-technological side, Franklin is best known for applying his knowledge of electricity to the invention of the lightning rod. A word of explanation: lightning was a serious problem in cities in the eighteenth century, particularly cities built primarily from wood, as in America. These cities had no central water supply and rudimentary firefighting capability at best, so a lightning strike not only threatened the building it hit, but could start a fire that could destroy entire neighborhoods. Franklin noticed that when he generated sparks with his static electricity experiments, there was always a flash of light and a "pop" sound. Thinking about lightning, he realized that it, too, exhibited these phenomena—the visible strike itself, followed by thunder. This suggested to him that lightning might be an electrical phenomenon.

His lightning rod was a simple iron rod that reached above the top of a house and was driven into the ground. In essence, it created a kind of super-highway for electrical charges in a thundercloud to get into the ground without having to fight their way through the house, providing a very effective way of diverting the stroke. Today we achieve the same effect by putting short metal rods on tall buildings and power lines, and then connecting those rods to the ground with flexible metal cables.

The reception of Franklin's invention was interesting. In America and England it was accepted enthusiastically, although in England there were minor quibbles about some details in the design. In France, however, clerics railed against the device because, they argued, it thwarted the will of God. In modern language, they accused Franklin of "playing God"—a charge often leveled today at those working in biotechnology (see Chapter 10). Consequently, in

France men still went up into church steeples during lightning storms to ring the bells. This was supposed to prevent lightning strikes, but in reality it's really hard to imagine a worse place to be in a thunderstorm than a church steeple. Many men were killed before the lightning rod was accepted in France.

If Benjamin Franklin was the prototype of the inspired tinkerer, Charles Augustin de Coulomb represented the careful, laboratory oriented side of science. Independently wealthy, he was not prominent enough to attract the attention of the French Revolution—in fact, he was called to Paris in 1781 to participate in the development of what we now call the metric system of units. He contributed to a number of fields of engineering and, believe it or not, he is known to builders as a pioneer in the design of retaining walls.

His electrical experiments, however, will be the focus of our attention. In 1785 he published a series of papers in which he reported on his detailed experiments on the electrical force. Putting carefully measured charges on objects a known distance apart, he would measure the resulting force. We can summarize his results as follows:

> Between any two charged objects there is an attractive force between unlike charges and a repulsive force between like charges that is proportional to the product of the charges and inversely proportional to the square of the distance between them.

This result is known as Coulomb's Law. You may notice the similarity between it and the Law of Universal Gravitation on pp. 71–72. With this result, we have said everything that can be said about the behavior of static electricity.

Animal electricity

The discoveries being made by the electricians were not of interest only to scientists. Perhaps the greatest public interest centered around the notion that electricity was somehow responsible for the properties of living systems. A common notion at the time was that living matter was different from non-living because it was imbued with a mysterious "life force." In the next chapter we will see how biologists in the nineteenth century proved this theory, called vitalism, to be wrong, but in the latter part of the eighteenth century, it was still very much in vogue.

Experiments done by the Italian physician Luigi Galvani (1737–98) certainly seemed to support the connection between electricity and living things. In a series of famous experiments, he showed that amputated frogs' legs could be made to twitch if they were stimulated by electrical charges. Later, Galvani showed you could get the same effect by touching the frog's legs with two different metals—copper and iron, for example. He argued that this showed that there was a special kind of force operating in the muscles, naming this vital force "animal electricity."

The Italian physicist Alessandro Volta, whose work will be discussed more fully below, engaged in a long debate with Galvani about the meaning of the experiments. Volta felt that what Galvani was seeing was the result of a chemical reaction between the fluids in the frog's leg and the metal probes. In the end, each of the men had half the answer. Galvani was right that muscle reactions can be stimulated by electrical current, as anyone who has ever touched an open circuit can testify, and Volta was right that such a current could be generated by chemical reactions.

In the end, then, there was no animal electricity—just ordinary positive and negative charges. The debate on the subject, however, had a number of surprising outcomes. One, discussed below, is that the investigations Volta was led to do inspired him to invent the battery, and this device, indirectly, led to the discoveries of the laws of electromagnetism that make modern society possible.

On a more exotic note, some researchers in the nineteenth century used batteries to study the effect of electrical current on human cadavers. They even gave public shows in which cadavers were made to kick their legs and sit up. Some scholars claim that Mary Shelley saw such a show and that it inspired her famous novel *Frankenstein*. If that is the case, the science of electricity had a long reach indeed.

Magnestism

The historical story of magnetism is a little different from that of electricity, primarily because magnets can be used as compasses. All of the ancient civilizations were aware that certain naturally occurring iron minerals, called lodestones, had the ability to attract pieces of metal to them. The Greeks even had a legend that somewhere in the Aegean there was an island made entirely of this material. They used this legend as a way of justifying the practice of not using iron nails in shipbuilding, since those nails would be pulled out if the ship ventured too close to the island. (There is, of course, no such island, but iron nails corrode when exposed to sea water—reason enough not to use them in ships.)

It quickly became apparent that there was a structure to magnetic materials. In modern language, every magnet has two poles, which we customarily label as "north" and "south." If the north poles of two magnets are brought near each other, they repel, while if the north pole of one magnet is brought near the south pole of another, they are attracted. This seems to be the basic law of magnetic force.

The important property of magnets is that if they are free to move—if, for example, a lodestone is placed on a cork floating in water, the magnet will line up in a north–south direction. Thus, the magnet can be made into a compass which navigators can use to determine direction even when they can't see the sun or the stars.

The question of where the compass originated and how the technology moved around in the ancient world engenders considerable debate among historians. Based on the discovery of an artifact made from lodestones in Central America, some scholars have suggested that the Olmec had developed the compass as early as 1000 BCE, although other scholars dismiss the claim, arguing that the artifact is simply part of a decorative structure.

Evidence for the use of the compass by the Chinese is unambiguous. It appears that the instrument was first used to determine orientations of buildings as part of the ancient art of *feng shui*. By 1040 CE, there are references to the use of magnets as "direction finders" in the Chinese literature, and most scholars accept this as the date for the invention. The first reference to the use of the compass in navigation appears in Chinese sources in 1117.

The question of how the compass came to Europe is also the subject of debate. We have references to the navigational use of compasses in European texts as early as 1190. Whether the invention somehow was brought to Europe from China, or whether this is yet another example of independent development remains an open question. It is clear, though, that by the high Middle Ages, European navigators had working compasses at their disposal.

The growing knowledge of magnets was codified and expanded by the English physician William Gilbert (1544–1603). His 1600 book *De Magnete* summarized both the accumulated knowledge about magnets and his own research. He was the first to realize that the Earth itself was a giant magnet, and it was the interaction between the Earth magnet and the north and south poles on bits of lodestone that made the lodestone line up north–south. Gilbert even made a small globe out of magnetic material to investigate the behavior of compasses.

He also realized another important fact about magnets, something that makes magnetic poles different from electrical charges. It is possible to have isolated electrical charges—for example, you can remove an electron from an atom to create an isolated negative charge. You can't do that with magnetic poles. If you take a large magnet with a north pole and a south pole and cut it in half, you do not get an isolated set of poles—you get two shorter magnets, each of which has a north and south pole. This phenomenon repeats indefinitely if you keep cutting up the smaller and smaller magnets. You simply cannot produce an isolated north or south pole.

So by the end of the eighteenth century, the study of both static electrical charges and magnets led to two general rules, which we write below for future reference

Coulomb's Law for electrical charges
There are no isolated magnetic poles

The most important fact from our point of view, however, was that there was no connection between these two sets of phenomena—they were simply two independent areas of study.

Fittingly enough, in 1800, the transition point into a new century, an important development occurred. The Italian physicist Alessandro Volta (1745–1827), after whom the Volt is named, produced the first prototype of what he called a voltaic pile and we would call a battery. For the first time, scientists had access to a device that could generate moving electrical charges—what we call today an electrical current. This opened a new chapter in the studies of electrical and magnetic phenomenon.

The first step along the new road occurred on April 21, 1820, when a young Danish professor named Hans Christian Oersted (pronounced ER-sted) (1777–1851) was lecturing in Copenhagen. Among the pieces of apparatus on the demonstration table were a battery and a compass. He noticed that whenever he connected the battery so that the electrical current flowed in the wire, the compass needle would move. What Oersted discovered that day was that there is, indeed, a connection between electricity and magnetism—that moving electrical charges (electrical current) can produce magnetic effects—today, we would say that they produce magnetic fields. These magnetic fields, in turn, are what make the compass needle move.

With this discovery, Oersted demolished the barrier between electric and magnetic phenomena. Together with the work of Faraday described below, it established that these two fields of study, so seemingly disparate, are just two sides of the same coin. In addition, although we will not have space to go into details here, the discovery led to important technological innovations such as the electromagnet and the electric motor, both of which are ubiquitous in modern society. This marks the first time, then, that we see a clear connection between basic research—research carried out solely for the acquisition of knowledge—and important economic benefits to society.

Michael Faraday (1791–1867) was a man who rose from humble beginnings to become the most prominent scientist in Victorian England, a frequent guest at Queen Victoria's court. The son of a blacksmith, he was, like Dalton, excluded from the English educational system because he was a religious dissenter. He was apprenticed to a bookbinder at the age of 14 and taught himself science by reading books as he was binding them. He attended some public lectures given by Humphrey Davy, a prominent chemist, took notes, and presented them to Davy bound in leather. This eventually led to Davy hiring Faraday and opened the scientific world to the brilliant young man.

Faraday made so many contributions to science that it's hard to know where to start describing them. If you will excuse a sports analogy, he was not only Hall of Fame, he was All Time All Stars. For our purposes, we will concentrate on one of these achievements—electromagnetic induction.

Here's a simple example to illustrate this phenomenon: imagine you have a loop of wire laying on a table—no batteries, no plug-in to a wall socket, just a loop of wire. Now imagine that I have a magnet in my hand, and I move it toward the loop. What Faraday discovered was that so long as the magnetic field in the region of the loop was changing, an electrical current would flow in the wire, even in the absence of batteries or other sources of power. We say that the changing magnetic field *induces* an electric current.

The most important device that came from the discovery of induction is the electrical generator, which produces virtually all of the electrical power used by modern civilizations. To visualize a generator, go back to our example of a loop of wire and a magnet, only this time let the loop be held up by a straight rod going across it and let the magnet be fixed in space. If you grab the rod and twist it, the loop will rotate. This means that the magnetic field *as seen by someone standing on the loop* will keep changing so long as the loop is rotating. Thus, so long as you keep turning the rod, an electrical current will flow in the loop—first one way, then the other. This is the principle of the electrical generator.

I can't leave Faraday without recounting a story. One of the English Prime Minister—probably Palmerston—was touring Faraday's lab and was shown a prototype of the generator.

"This is all very interesting, Mr. Faraday" he said, "but what good is it?"
"Mr. Prime Minister," Faraday replied, "someday you'll be able to tax it!"

Maxwell's Equations

When the 'electricians' of the nineteenth century were finished, then, there were four basic laws governing electricity and magnetism:

Coulomb's Law.
There are no isolated magnetic poles.
Electrical currents produce magnetic fields.
Changing magnetic fields produce electrical currents.

These four statements, in suitable mathematical form, are called Maxwell's Equations, and they play the same role in electricity and magnetism that Newton's Laws play in the science of motion. Every phenomenon involving electricity and magnetism, from the interactions of electrons in atoms to the magnetic fields of galaxies, is governed by these laws. They form an important pillar of what is called classical physics (that is, the physics that developed before the twentieth century).

You may be wondering why the equations are named after a man who had nothing to do with discovering them. James Clerk Maxwell (1831–79) was a Scottish physicist who made contributions to many fields of science. As far as the equations are concerned, he added a small term to the third equation (it's really of interest only to experts) that completed the system. Second, he was well versed in the forefront mathematics of his time (which, in those days, was partial differential equations), and this skill allowed him to see that these particular equations form a coherent, self-contained whole, and also allowed him to manipulate the equations to produce important new predictions.

Once he had the equations in order, Maxwell showed that they made a startling prediction. According to Maxwell's calculations, there should be waves in nature—waves that could move through empty space by tossing energy back and forth between electric and magnetic fields. Furthermore, the speed at which these waves moved was predicted to be 186,000 miles per second—the speed of light. In fact, Maxwell quickly identified visible light, from red to violet, as an example of these new "electromagnetic" waves. The problem was that his theory predicted that these new waves could have any wavelength, while visible light has wavelength from around 4,000 atoms across (for blue light) to 8,000 atoms across (for red light). In fact, the rest of the waves that Maxwell predicted were quickly discovered, from radio waves to X rays. The existence of this so-called electromagnetic spectrum was a confirmation of the validity of Maxwell's equations, of course, but it also provided us with an enormous range of useful functions—you use various parts of the spectrum, for example, every time you use a cell phone or warm food in a microwave.

The electrification of America

The point about the electrical generator is that as long as you can find a source of energy to spin the coil of wire in a magnetic field, you can produce electrical current. That current can then be sent over wires to distant locations, so for the first time in history people were able to generate their energy in one place and use it in another—an important prerequisite for the rapid growth of urban areas that has characterized the past century. Today, the energy to spin the coil can come from falling water in a dam, but more likely will come from a jet of steam generated by a coal or gas boiler or a nuclear reactor.

The first commercial demonstration of the new technology took place in Chicago on April 25, 1878, when a series of carbon arc lamps were used to light up Michigan Avenue. The demonstration was successful, but the second night the whole system blew up and couldn't be reinstated for months.

Despite this inauspicious beginning, electrification caught on quickly. At first, large private institutions like hotels and department stores installed their own generators, replacing the existing gas lamps with electric lights. By 1879, the first central generating station had been built (in San Francisco) and others soon followed in other cities. At first, the limiting factor was the cost of the copper wire to carry the current from the generator to the user—at one point, for example, it was uneconomical to transmit the current more than 16 blocks from the generator. As the technology advanced and old fashioned nickel-and-dime engineering improved the system, these barriers were overcome, and our current system of massive central power plants linked to a ubiquitous grid began to take shape. With the creation of the Rural Electrification Administration in 1936, the United States became the first country to create a continent-wide electrical grid.

The point of the story is this: if you had asked Faraday or Maxwell whether their work would lead to better street lighting or the electric toothbrush, they would have had no idea what you were talking about. They were interested in discovering the basic laws of nature—in doing what we would call basic research today. Yet within a few decades of their work, the complex process of wiring the planet was well underway. This sort of thing is often the case with basic research—it leads to places that no one can predict in advance.

When I discuss this with my students, I often put it in terms of a notion to which I have given the tongue-in-cheek name of "Trefil's Law." It states:

> Whenever somebody discovers something about how the universe works, pretty soon someone else will come along and figure out how to make money from the discovery.

Summary

One way of gauging the impact of science and technology in the nineteenth century is to compare the life of someone in 1800 with a similar life in 1900. The differences are startling.

- In 1800, if you wanted to travel from one place to another, you used the same type of transportation—human or animal muscle power—that was available to the ancient Egyptians. In 1900, thanks to the inspired tinkering of many individuals, you would take a train, harnessing the solar energy stored in coal.
- In 1800, if you wanted to communicate with a friend across the ocean, you would write a letter that could take months to reach its destination. In 1900, the invention of the telegraph and the laying of the transatlantic cable reduced the communication time to minutes.
- In 1800, if you wanted light at night you burned candles or used whale oil lamps. By 1900, you probably flipped a switch and turned on an incandescent light bulb, powered by electricity.

I could go on, but I think the point is clear. Throughout the nineteenth century science and technology interacted in many different ways to produce enormous changes in the human condition. The steam engine, as we saw in Chapter 1, was the product of engineering work done by James Watt. In this case, the basic science (a branch of physics called thermodynamics) was developed later, at least in part because of the practical importance of the engine. The electrical generator, on the other hand, grew out of Faraday's investigations into the basic properties of electricity and magnetism—in this case, the basic science led the technology.

The story of the telegraph is a strange mix of science and technology. Its operation depends on the use of electromagnets, which, as we have seen above, resulted from Oersted's work on the basic properties of electricity and magnetism. The first working telegraphs—in England in 1836 and America in 1838—relied on engineering improvements in these devices. Consequently, we can think of those telegraphs as resulting from a combination of basic science and technology, with technology taking a slight lead.

The transatlantic telegraph cable, however, is a somewhat different story. The first of these cables to be laid successfully in 1866 extended from Ireland to Newfoundland. The operation of the cable depended on the work of British physicist William Thomson (later Lord Kelvin) (1824–1907) who used the newly discovered laws of electromagnetism to work out the theory of the coaxial cable, eventually showing how faint telegraph signals could be sent and received over long distances. At this point, it would appear that the basic science was leading the technology.

Inventions like the telegraph can affect people's lives in strange and unexpected ways. Let me illustrate this with one example. Before the telegraph, there was no way for people in Boston (for example) to know what the weather was like in New York or Philadelphia in real time. The concept of something like a weather front or a widespread storm simply could not be imagined. Once someone could gather data by telegraph from many places, however, it became possible to see weather patterns as a geographical phenomenon and to begin making predictions. You can think of the telegraph, then, as a kind of precursor to modern weather satellites and perhaps even to the Weather Channel.

Chapter 8

The science of life

It is no accident that biology, the branch of science devoted to studying living systems, lagged behind the physical sciences in development. The fact of the matter is that living systems are a lot more complicated than non-living ones. It can be argued that a single amoeba, for example, is more complex than an entire galaxy, and that the human brain, composed of billions of inter-connected neurons, is the most complex system in the universe. Small wonder, then, that the science of life was slower to develop. Like the physical sciences, however, biology underwent a profound transformation in the post-Newtonian world. Basically, it went from being a science devoted to describing and cata-loguing living things to a science concerned with the working of cells as the basic unit of living systems. I will call this the "cellular turn," and, as we shall see, it was a precursor to the emphasis on the interactions of molecules in today's biology. Unlike the events surrounding Isaac Newton's work, how-ever, the shift in attention to cells was gradual, with no single person leading the transition. In what follows, then, I will pick scientists who illustrate this new way of thinking and ask you to keep in mind that the final product was the work of many scientists who we don't have space to discuss.

As we have done in previous chapters, we will trace the cellular turn by looking at the lives of a few of the principal scientists involved. These men are:

Carl Linnaeus (Sweden—1707–78)
Gregor Mendel (Austria—1822–84)
Louis Pasteur (France—1822–95)
Theodor Schwann (1819–82) and Matthais Schleiden (1809–81), both German.

To this list we will add:

Charles Darwin (England—1809–82).

There is a reason for keeping Darwin separate. Arguably the greatest biol-ogist who ever lived, he was not part of the cellular turn. Instead of thinking about how living systems worked, he asked a different question. He wanted to understand how living things came to be the way they are, and this introduced a whole new mode of thought into the life sciences.

Carl Linnaeus

For most of recorded history, there was a single goal for biologists—to describe and classify all of the living things on our planet. This is truly a daunting task, as you can see by noting that within a few hundred yards of where you are sitting right now, there are at least hundreds, and probably thousands, of different species of plants and animals. Today the best estimates of the total number of species on Earth run to around 8 million, a daunting assortment to classify without the aid of computers. Carl Linnaeus represents the culmination of what we can call the "identifying and classifying" phase of biology.

Born in rural Sweden, he was educated and spent most of his academic career at the University of Uppsala, where he studied both medicine and botany and eventually became rector. He is remembered because he created a way of organizing the immense diversity of living systems that, with modern modifications, is still in widespread use today.

Next time you go to a zoo or museum, pay close attention to the placards introducing the animals. You will notice that below the English title, there will be two words in Latin—*Ursus maritimus* for the polar bear, for example. This so-called "binomial nomenclature" was introduced by Linnaeus as part of his classification scheme. The scheme itself is based on a sorting process, in which plants or animals with similar properties are grouped together, and dissimilar organisms are separated. For example, you might start classifying a squirrel by noting that a squirrel is more like a snake than it is like an oak tree, then noting that a squirrel is more like a rabbit than it is like a snake, and so on. As an example of how the system works in its modern form, let's see how a contemporary Linnaeus might go about classifying human beings.

For starters, he would want to put human beings into a large class known as a kingdom. Linnaeus recognized two kingdoms—plants and animals—but today we usually add three more—fungi and two kingdoms of single-celled organisms. Humans, who ingest their food, are clearly in the animal kingdom. We are more like snakes than oak trees.

Among animals, there is a group that has a nerve cord running down the back, usually encased in bone. Humans have a backbone, so we are in the so-called "phylum" that includes vertebrates. We are more like snakes than we are like lobsters. Among vertebrates, there is a group that is warm blooded, gives live birth, and nurses its young. Humans are in the "class" of mammals— we are more like rabbits than we are like snakes. Among mammals there is a group with prehensile fingers and toes and binocular vision. Humans are clearly in the "order" of primates—we're more like monkeys than we are like rabbits.

At this point in the classification we run into something mildly unusual about humans. Most animals have a large number of close relatives. We do, too, but all of them happen to be extinct. For the record, we know of some 20-odd extinct "humans."

To resume our classification, we note that among primates there is a group that walks upright and has large brains. Humans belong to the "family" of hominids—we are more like Neanderthal man than we are like monkeys. At this point, the criteria for distinction become very detailed—things like tooth shape or sinus cavities become the focus of attention. But among hominids, humans belong to the genus *Homo*—we are more like Neanderthal than we are like other fossil hominids. Finally, we get to our specific species, which characterizes us and no one else—the species *sapiens*. Thus, the binomial classification for humans is *Homo sapiens* ("Man the wise").

To summarize the Linnean classification for humans:

Kingdom—animal
Phylum—chordate (most chordates are also vertebrates).
Class—mammals
Order—primates
Family—hominids
Genus—*Homo*
Species—*sapiens*

There is, incidentally, a nice mnemonic for remembering the Linnaean categories—"King Phillip came over for good spaghetti." We could, in principle, go through the same sort of process for any of the 8 million species on our planet, from frogs to butterflies. In the end each would fit into its own unique pigeonhole. Such an outcome would indeed represent the culmination of the "describe and classify" stage of biology.

Having said this, I have to point out that there is nothing sacred or inevitable about classifying organisms as Linnaeus did. In fact, in Chapter 10 we will see that the final outcome of the "cellular turn" in the nineteenth century was a "molecular turn" in the twentieth, when biologists began to pay attention to the molecular processes that govern the function of cells. In this case, many scientists find a classification scheme based on molecular structures more useful than the scheme based on the gross structure that Linnaeus used. What matters with classification is usefulness, and there are many possible ways of organizing living systems.

Gregor Mendel

The stereotype of the scientist as a lonely genius working in isolation should be seen by this point in our development to be a poor description of the way that science actually moves forward. Every scientist we've considered has had colleagues with whom he could discuss his work and predecessors upon whom his work could be based. Gregor Mendel is, in fact, the only scientist we will discuss who fits the stereotype of the lonely genius.

Mendel was born in what is now the Czech Republic, and after a period studying in Vienna, he joined the Augustinian Abbey of St. Thomas in Brno as a teacher of physics. In 1867, he became abbot of the monastery. Between 1856 and 1863, however, Mendel performed a series of experiments that would, eventually, change the way that scientists think about living systems.

The region around the monastery at Brno is dominated by agricultural pursuits, with rolling hills, orchards, and vineyards. It is, therefore, not surprising that Mendel took as the object of his scientific study the problem of inheritance—the question of how specific traits of parents are passed on to offspring. There was, of course, a great deal of practical knowledge on this subject, accumulated by millennia of plant and animal breeding. Mendel chose the standard garden pea as his model for exploring this question.

A word of explanation: Mendel was not really interested in examining the genetics of peas. He chose the plant because in the climate of central Europe a gardener can raise three crops of peas to maturity in a single growing season. This means that peas are an ideal plant for what biologists call a "model." The idea is that if you can understand something about the inheritance rules for peas, where you can see several generations in a few years, you have a pretty good idea about the inheritance rules for apple trees, where a generation may last a decade or more.

A classic Mendelian experiment would go like this: he would start by obtaining plants that were "true-breeding"—that is, plants where the offspring have the same traits as the parents. For example. If the parents of a true breeding pea plant are tall, the offspring will also be tall. Mendel would take the pollen from (for example) tall plants and fertilize short ones, then wait to see how the offspring turned out.

He quickly discovered that there were repeatable patterns in inheritance. For example, in the experiment described above, all of the offspring of the first cross will be tall. If, however, we do that same sort of cross fertilization on those offspring, only ¾ of the second generation of plants will be tall, and the remaining ¼ will be short. Similar rules came from other experiments.

Mendel realized that his data could be understood if there was a "unit of inheritance," which he called a "gene," and if each offspring got one such unit from each parent. If both parents contributed a gene that gave the same instruction—for example, "grow tall"—there was no problem. If the instructions from the genes were conflicting for example, "grow tall" from one parent and "grow short" from the other—then one of the genes would "win." This gene is said to be "dominant." In the example given above, for example, the gene for tallness is dominant, since all the first generation plants are tall, even though each of them received a gene for shortness from one parent. The short gene is said to be "recessive." It is carried along, but it is not, in the language of biologists, "expressed." Thus, each first generation plant carried both a "tall" and a "short" gene, which we can represent as (Ts). If we cross the first generation plants, then the possible gene combinations in the second generation will be:

(TT)
(Ts)
(sT)
(ss).

Of these four possible gene combinations, the first three will produce tall plants, and only the last will produce a short one. Thus, the assumption that inheritance is governed by genes that obey the rules given above explains the experimental observation that ¾ of second generation plants are tall. A similar explanation can be produced for the rest of Mendel's results.

Mendel presented his results to the Brno Natural History Society, and actually published them in the Society's journal. What followed was one of the strangest sequences of events in the history of science. The journals languished, unread, in European libraries for almost half a century while Mendel devoted the remainder of his life to running the monastery. In the late nineteenth century, two scientists investigating inheritance—the Dutch geneticist Hugo de Vries and the German botanist Carl Correns—not only duplicated some of Mendel's original results independently, but found his original papers and called them to the attention of the scientific community. Largely because of these men's personal integrity, we now understand the role that Mendel played in our knowledge of the operation of living systems.

One more point about Mendelian genetics. For Mendel, the gene was an intellectual construct—something that helped explain his findings. For us, as we shall see in Chapter 10, it is something different. Today, we see the gene as a real physical object—a stretch of atoms along a DNA molecule. In a sense, this transition from construct to concrete object is the story of-twentieth-century biology.

The cellular turn

As we said above, the shift of attention away from organisms to cells was a gradual one, involving many different scientists over several centuries. The British scientist Robert Hooke (1635–1703), a contemporary of Isaac Newton, first identified cells using a primitive microscope to examine slices of cork in 1665. Apparently, he saw the cork broken up into small compartments that reminded him of the cells monks occupied in monasteries, and this led him to give the structures their current name. The first living cells were seen under a microscope in 1674, when the Dutch scientist Anthony van Leeuwenhoek (1632–1723) saw moving "animalcules" in a drop of water. The

French physician and physiologist Henri Dutrochet (1776–1842) is generally credited with pointing out that the cell is the fundamental element in all living systems.

Insofar as we can assign a symbolic date to the start of the cellular turn, it would be a dinner in 1837 between Matthias Schleiden and Theodor Schwann. Schleiden had been educated as a lawyer, but his interest in botany quickly won him a professorship at the University of Jena. Schwann had been trained in medicine, but had moved into a research career as an assistant at a museum in Berlin (he would later hold many prestigious academic positions). Both men had done extensive microscopic studies of cell—Schleiden of plant cells, Schwann of animals. In their dinner conversation, they realized that there were important similarities between the two, a realization that quickly resulted in a publication which declared that "All living things are composed of cells and cell products that are reproduced." With the later addition of the statement by Rudolph Virchow (1821–1902), that all cells are produced by the division of pre-existing cells, this statement established the basics of cell theory.

It's difficult to overstate the importance of this development. For the first time in history scientists were looking at living systems as manifestations of known physical and chemical laws, rather than as the expression of some mysterious "life force." When Louis Pasteur's experiments (described below) established definitively that life did not arise by spontaneous generation, but from pre-existing life, the basis was laid for treating biology like any other branch of science. As we shall see in Chapter 10, there is a direct link between this idea and our current understanding of the molecular basis of life.

Louis Pasteur

No scientist personifies the cellular turn better than Louis Pasteur, a fact that posterity has honored in many ways. All over the world streets, buildings, and research institutes are named in his honor. To top this list, we note that in the last episode of *Star Trek: The Next Generation* the Starship U.S.S. *Pasteur* plays a prominent role. What did the man do to deserve this kind of recognition?

Louis Pasteur was born into a working class family in southeastern France, but his talents were quickly recognized, and he received the best education that the French system could offer. After a brief period studying crystals, Pasteur turned his attention to what is now called the germ theory of disease—arguably his greatest contribution to future generations.

In the mid-nineteenth century, the theory of spontaneous generation— the idea that life could arise spontaneously from non-living matter—was still very much alive. There had been a long history of experiments designed to debunk the idea, starting in 1668. In that year Francesco Redi showed that if flies were kept away from rotting meat (by putting the meat under gauze or in a bottle, for example), no maggots would appear. This showed that the maggots were not generated spontaneously by the meat, but came from eggs laid by flies. The evidence against the theory of spontaneous generation was fragmentary, however, and it remained for Pasteur, in 1859, to provide the final comprehensive evidence. He put broths that had been thoroughly boiled into flasks that either had filters or long curving necks to prevent the entry of dust. He found that nothing grew in the isolated broths, whereas broths exposed to the air were quickly contaminated.

Pasteur concluded that the contamination of the broths was caused by organisms in the air, probably riding along on microscopic dust particles. It wasn't the broth that generated the contamination, in other words, but (in modern language) microbes brought in from the outside. And if this was true for broths, it's not much of a step to see that it would be true for infections and communicable diseases in humans. This is, in essence, the germ theory of disease. And although Pasteur wasn't the first to suggest the theory, he was the first to have hard experimental evidence to back it up.

The societal impact of his work was immediate and enormous. In 1862, with his colleague Claude Bernard (1813–78), he demonstrated that microbes in milk could be killed if the milk were heated. This process, known as pasteurization, is now used universally to make dairy products safe to consume. Pasteur also suggested that many diseases could be prevented if microbes were not allowed to enter the human body. After reading Pasteur's papers, the British surgeon Joseph Lister (1827–1912) began experimenting with using chemical antiseptics to prevent infection in surgery. He published his results in 1867, and his techniques were widely copied. The concept of the sterile operating room, so much a part of modern medicine, is thus another consequence of Pasteur's work on the germ theory of disease.

In later life, Pasteur worked in the field that would be called immunology today. He discovered the process of immunization through the use of weakened bacteria by accident. He was studying the transmission of chicken cholera, and had told an assistant to inoculate some chickens with a sample of bacteria while he (Pasteur) went on vacation. The assistant failed to do so, and the bacteria started to die out. When chickens were inoculated with these weakened samples they got sick, but didn't die. Afterwards, inoculation with full strength bacteria could not produce the disease in those particular chickens— today, we realize that the first inoculation triggered an immune response that allowed them to fight off full scale infections.

Although this technique had been known to work for smallpox, Pasteur quickly applied it to anthrax and rabies. Next time your doctor gives you a shot to keep your immune protection current, you can say a silent thanks to the great French scientist.

What Pasteur did, then, was to take the cellular turn very seriously and view the world as populated by unseen organisms capable of causing disease and decay. From this insight flowed all of the benefits we've just discussed, and which we now take for granted. Clearly, the man deserves to have a Starship named in his honor!

Charles Darwin and evolution

Charles Darwin was born into a prominent and wealthy family. His grandfather, Erasmus Darwin, had been a noted physician and philosopher, and his father was a physician as well. Young Charles was sent off to medical school in Edinburgh, but it turned out he was squeamish about the sight of blood— then, as now, a pretty good argument against a career in medicine. His father then sent him to Cambridge University to study theology, in the hope that he would pursue a career as a minister. At the university, however, Darwin made the acquaintance of the botanist and naturalist John Henslow, and quickly became interested in studying the natural world.

In those days Great Britain was the pre-eminent maritime power in the world, and was heavily involved in exploring the world's oceans. Henslow was offered a position as naturalist on a ship named the *Beagle* which was scheduled to make an extended voyage around the southern hemisphere. He decided not to accept the offer, but suggested that Darwin, recently graduated, would be an excellent candidate for the post. Thus began one of the most important voyages of discovery in the history of science.

The *Beagle* sailed around South America while Darwin collected geological and biological specimens. And although it would not become obvious until later, the crucial port of call turned out to be the Galapagos Islands, a thousand miles west of Peru. In fact, "Darwin's finches" have joined "Newton's apple" in the folklore of science as seemingly ordinary events that led to extraordinary advances in science.

Some background: in the early nineteenth century, the standard explanation for the diversity of biological forms in the world was that each species had been specially created by God in its present form and had not changed over the (relatively brief) history of the Earth. This is what Darwin had been taught. Yet here in the Galapagos he found all these finches that were obviously closely related, but which were clearly not the same. Why, Darwin wondered, would the Creator go through all the trouble of making dozens of species that were so alike, yet so different?

After his return to England, Darwin spent decades developing his ideas and establishing his reputation as a naturalist through the study of organisms like barnacles and earthworms. During this period he slowly developed the ideas about natural selection discussed below. He planned to write a multivolume work explaining the new ideas he was developing about the natural world. It was quite a shock, then, when in 1858 he received a letter from a young naturalist by the name of Henry Russell Wallace in what is now called Indonesia. In essence, Wallace was proposing the same idea about natural selection that Darwin had developed through decades of research. Darwin quickly wrote a letter summarizing his views and had both his and Wallace's letters read at a meeting of the Royal Society, then set to work on writing what he undoubtedly regarded as a quick summary of his ideas. The result was the 1859 publication of *The Origin of Species*, arguably the most influential scientific book ever written.

Darwin begins his argument by talking about the well known fact that plant and animal breeders have known for millennia that it is possible to improve the characteristics of a breed by careful choice of parents. A pigeon breeder himself, he talks about how, in a relatively short period of time, breeders have produced an astonishing variety of birds. This process, in which human beings, by conscious choice, change the characteristics of a breed is called "artificial selection." The fact that it works as well as it does implies that there has to be a mechanism by which the characteristics of the parents are passed on to offspring.

It is important to understand that, while Darwin and his contemporaries were aware that such a mechanism had to exist, they had no idea what the mechanism actually was. This is ironic because, as we pointed out above, the work of Mendel on genes was already sitting, unnoticed, in their libraries. Although critics faulted Darwin for his failure to explain this mechanism, the development of evolutionary theory really didn't require knowing what the mechanism of inheritance is—you just had to know that it exists.

With this background, Darwin proceeded to ask one of those questions that is obvious in retrospect, but requires a genius to ask the first time. Was it possible, he asked, that there was a mechanism in nature that could, without human intervention or intent, produce the same sorts of results that artificial selection does? Does there exist, in other words, a mechanism we could call "natural selection"?

Darwin's positive answer to this question depends on two simple facts about nature:

1   there is variation within populations
2   individuals must compete for scarce resources.

In a given population, abilities and characteristics will vary from one individual to the next. Today, we know that this has to do with genes and DNA, but for Darwin's purposes all that mattered was that the differences obviously existed. Some animals could run faster than others, some had more fur, some had more advantageous coloring, and so on. Depending on the environment, any of these characteristics might convey a competitive advantage, allowing its carrier to live long enough to reproduce and pass the characteristic on to its offspring. (Note that in modern language, we would talk about passing on genes.) For example, a rabbit that can run fast would have a survival edge in an environment that contained predators. Thus, in modern language, the next generation of rabbits will have more "fast" genes than the first one, simply because more fast rabbits survive long enough to reproduce. This process will go on from one generation to the next, with each succeeding generation having more individuals with those genes than the previous one. This, in a nutshell, is natural selection. In Darwin's view, small changes eventually accumulated to produce different species—hence the title of the book.

There are a couple of points that need to be made about *Origin*. The first is that, as mentioned above, Darwin's argument was lacking a crucial ingredient—an understanding of how traits are passed from one generation to the next. As we shall see in Chapter 10, it would be over half a century before Mendelian genetics and modern molecular biology were combined with Darwinian evolution to produce the modern version of evolutionary theory.

Second, Darwin consciously avoided including the evolution of human beings in his discussion. This was partly because he realized that including humans in the evolutionary framework would bring him into conflict with the religious establishment of his time, and partly because he was concerned not to disturb his wife, who was a devout Christian. It wasn't, in fact, until 1871 that he bit the bullet and published *Descent of Man*, including humans in the evolutionary scheme. This inclusion, of course, has been the main point of contention between scientists and religious fundamentalists in the United States ever since.

One thing is clear, however. In its modern form, the theory of evolution lies at the base of the life sciences and medicine. It ties these fields together conceptually and provides a kind of intellectual superstructure for them. Without evolution, as many scientists have noted, nothing in the life sciences makes sense. And although there may be arguments among scientists over specific details of the theory, there is no argument about whether it provides a reliable guide to the development of life on our planet.

Reactions to Darwin

As was the case with Copernicus, the revolutionary concepts introduced by Darwin generated a long and often rancorous debate. We can identify two major areas of conflict. First there are both historical and contemporary examples of people misinterpreting Darwin to further a particular agenda, and, second, there are the conflicts with organized religion alluded to above. Let's look at these separately.

Probably the most egregious misunderstanding of Darwin was the rise of theory that acquired the misleading title of "Social Darwinism." Most often identified with the English engineer and philosopher Herbert Spencer (1820–1903), the basic idea of this philosophy was to apply Darwin's "survival of the fittest" (a phrase Darwin never used until Spencer popularized it) to Victorian social structure. The basic premises of Social Darwinism (all incorrect) were that

1   evolution implied progress
2   natural selection applied to societies
3   the "lower social orders" represent earlier stages of evolution.

The "fittest," those at the top of society, were there because in some sense they
had won the evolutionary battle, and Social Darwinists argued that it was
pointless to try to change this fact through government action.

It's really hard to know where to start criticizing an argument that is wrong
in so many ways. At the most basic level, in Darwin's theory "fitness" has to
do with success in getting genes into the next generation, not with money or
social position. It applies to individuals (or, more exactly, to their genes)
and not to groups or societies. Thus, the last two tenets above are simply
misinterpretations of Darwin.

The notion of "progress" is a little harder to talk about and, indeed, you
still run across it in some modern writings. Had you been born in the mid-
nineteenth century and seen the growth of technologies like the railroad, the
telegraph, and the electric light, it's not hard to see how you might believe in
the inevitability of progress. The chastening lessons of the twentieth century,
when men like Joseph Stalin and Adolph Hitler showed that technological
advances do not necessarily lead to human welfare, were still far in the future.
In Spencer's time, you might say that progress was "in the air."

Fair enough, but was it in Darwin? A moment's thought should convince
you that there is absolutely no implication of "progress" in Darwin's theory.
Organisms respond to an ever changing environment, with no goal or end in
mind. All that matters is surviving long enough to pass the genes on. The fact
(for example) that there are some 20-odd species of "humans" that became
extinct in the past and only one (us) made it to the present is evidence that the
evolutionary game is much more like a crap shoot that a smooth upward
slope. Examine the fossil record of almost any species and you'll see the same
thing—every blind alley stumbled into until something finally works.

The modern misinterpretations of Darwin are both less egregious and less
harmful than was Social Darwinism. Mainly found in popular magazines,
these analyses usually start with a naïve interpretation of some Darwinian
principle and try to draw lessons for human behavior from it. Hence we get
evolutionary dieting, evolutionary psychology, and even (God help us!) evolutionary marriage counseling. Most of these approaches commit what philosophers call the "is–ought fallacy." The fallacy lies in the fact that discovering
that nature behaves in a certain way does not imply that humans should do
likewise. Some male monkeys, for example, kill the offspring of other males to
increase their own reproductive success. Needless to say, no one argues that
humans should do the same.

The religious issues that Darwin stirred up revolve around the fact that the
major western religions—Christianity and Judaism—have a creation story in
the Book of Genesis that doesn't look a lot like *The Origin of Species*. Having

said that, I have to point out that most mainstream theologians of the Jewish, Catholic, and Protestant faiths long ago came to terms with Darwin. In the United States, however, a vociferous group of fundamentalist Protestants have kept the conflict alive, primarily by trying to introduce instruction in traditional creationist doctrine into the public schools. These attempts have invariably been challenged and defeated in the courts.

The central issue in all of these court cases is the First Amendment to the United States Constitution, which reads in part "Congress shall make no law respecting an establishment of religion." Courts have consistently ruled—most recently in 2005—that attempts to introduce Biblical narratives into school curricula are, in fact, nothing more than an attempt to introduce religious teachings under the guise of science.

Summary

If we perform the same exercise for the life sciences that we did for the physical, by comparing life in 1900 to life in 1800, we get a slightly different result. To be sure, there were important practical changes in everyday life due to advances in biology, most of them connected to the germ theory of disease and outlined below. More important, though, was a profound shift in the way that scientists looked at living systems. Instead of seeing them as something separate, different from the rest of the universe, they became legitimate objects of scientific study. This shift, as we shall see, has continued through to the twenty-first century and can be thought of as the driving force behind today's biotechnology revolution. Let's look at these two effects—one practical, one intellectual—separately.

We have already discussed several important products of the germ theory of disease—pasteurization, antiseptic surgery, vaccines. Even more important, however, was the way that the growing awareness of microbes influenced the way that people thought about what would today be called public health. In 1854, the British physician John Snow (1813–58) discovered that a cholera epidemic in London was caused by contaminated water in a public well. Up until that time, people had assumed that if water looked clear, it was fit to drink. The idea that there might be invisible organisms capable of causing disease eventually led to our modern system of water purification and waste treatment, a development that has saved countless human lives over the years.

The intellectual transformation of the nineteenth century came in two parts. One, which we have called the cellular turn, involved the realization that living things (including humans) were not different from other objects in the universe, and could be studied with the normal techniques of chemistry and physics. The second part, associated with Charles Darwin, was the understanding that the laws of natural selection could explain the historical development of living things on the planet. In many ways, this was the more difficult of the two transformations for people to accept, because it seemed to deny the notion that humans are, in some sense, special. As we mentioned above, most major religious thinkers have come to terms with Darwin by now, but the aimlessness and purposelessness of the evolutionary process is still a cause of concern for many.

Chapter 9

Physical sciences in the twentieth century

By the end of the nineteenth century, then, the scientific method had been applied to the study of many aspects of the physical world. In physics, for example, there were three thoroughly studied fields: mechanics (the science of motion), thermodynamics (the science of heat and energy) as well as electricity and magnetism. These three fields, collectively, are now referred to as "classical physics" and, taken together, they did a superb job of describing the everyday world that Isaac Newton had studied—the world of normal sized objects moving at normal speed.

As we saw in Chapter 7, the flowering of classical physics had enormous effects on the lives of ordinary people in the nineteenth century, bringing in everything from the telegraph to commercial electricity. In the same way, scientific advances in the twentieth century have changed forever the way that human beings live, communicate with each other, and generate energy. In this chapter we will look at advances in the physical sciences which led to things like the internet and nuclear power, while in the next chapter we will look at the parallel advances in the life sciences.

At the turn of the twentieth century, new experimental results were pointing to an entire new world to be explored—the world inside the atom. In addition, in 1905 an obscure clerk in the Swiss Patent Office by the name of Albert Einstein published an extraordinary paper that opened the door to yet another area that Isaac Newton had never thought about—the world of objects moving at speeds near that of light. When the dust had settled a few decades later, two new fields of science had been born—the theory of relativity and quantum mechanics, which describes the world of the atom.

It is common to talk of these two developments in the early twentieth century as "revolutions," which they were, and to suggest that in some sense they "replaced" Newtonian science, which they most assuredly did not. To understand this statement we have to go back to Chapter 1 and recall that all of science is based on observation and experiment. The entire edifice of classical physics was built on observations of the everyday world—what we have called normal sized objects moving at normal speeds. Classical physics works for the world which corresponds to its experimental base—this is why we still teach Newtonian mechanics to students who will go out and design bridges and aircraft.

The worlds of the atom and of high speeds, however, are not included in the Newtonian experiment base, and there is no reason to expect they should be governed by similar laws. In fact, we will find that both of these areas of the universe are governed by laws that, at first glance, seem very strange to us. This is because our intuition about how the world should work is based on our experience of the Newtonian world. In fact, most people's reaction when they encounter relativity and quantum mechanics for the first time is disbelief— "It can't be this way!" It takes a while to get used to the fact that there are parts of the universe that just don't look like our familiar world. It should be a consolation, however, to learn that if you apply the laws of relativity or quantum mechanics to billiard balls and other familiar objects, the new laws give us the good old Newtonian results.

As strange as these new laws may seem to you, however, you have to keep in mind that every time you turn on a computer or use a cell phone you are using the laws of quantum mechanics, and every time you use a GPS system (in you car's navigation system, for example) you are using the laws of relativity. These laws may indeed be strange, but they are already part of your life.

With this general introduction to the physical science of the twentieth century, then, let us turn to a discussion of the new sciences. For pedagogical reasons, we'll start with relativity.

The theory of relativity

The easiest way to introduce the concept of relativity is to think about a simple experiment. Imagine a friend is a passenger in a car going by on the street. As the car passes you, your friend throws a ball into the air and catches it when it falls. Your friend will see the ball go straight up and down, while you will see the ball traveling in an arc. If I asked the two of you to describe what you saw, then, you would give different descriptions of what the ball did. In the jargon of physics, your description of events would depend on your "frame of reference."

Suppose I asked a different question, however. Suppose that instead of asking you to describe the event, I asked you to find the laws of nature that govern the event. In this case, you would both come back to me with Newton's Laws of Motion. In other words, while the description of events may vary from one frame of reference to another, the laws of nature do not. This is called the Principle of Relativity, and is stated formally as:

The Laws of Nature are the Same in All Frames of Reference.

As it happens, the mathematical structure of Newton's Laws guarantees that this principle holds in the Newtonian world.

Before we go on, let me make an historical aside. The paper Einstein published in 1905 dealt only with frames of reference moving at constant velocities (i.e. frames that are not accelerating). If we restrict the principle in this way, it is referred to as the "Principle of Special Relativity." The principle as stated above, which applies to all frames of reference whether they are accelerated or not, is called the Principle of General Relativity. The fact is that it took Einstein over a decade to work his way from special to general relativity, which he published in 1916.

The problem Einstein addressed has to do with the fact that Maxwell's Equations predict the existence of electromagnetic waves, as we saw in Chapter 7. Part of this prediction was a statement about how fast these waves move—a quantity we customarily refer to as the "speed of light" and denote by the letter "c." Thus, the speed of light is built into the laws of nature, and if the principle of relativity is true, it must be the same in all frames of reference.

This statement is so crucial that it is often stated as a second postulate, along with the principle of relativity, when the theory is presented. The fact that the speed of light has to be the same in all frames of reference, however, leads to a fundamental problem. The easiest way to see this is to go back to your friend driving by in a car. Imagine that now, instead of a ball, your friend has a flashlight, and you both measure the speed of the light it emits. Your friend will get "c," of course, but, if the principle of relativity is correct, so will you—if you didn't, Maxwell's equations would be different in your frame of reference than they are in the car. At first glance, the fact that you will measure the speed as "c" and not as "c plus the speed of the car" just doesn't make sense.

There are three ways out of this dilemma:

1 The principle of relativity (and hence Newtonian mechanics) is wrong.
2 Maxwell's equations are wrong.
3 There is something wrong with our notion of velocity.

The first two options have been explored both theoretically and experimentally, and they turn out to be dead ends. Einstein looked at the third option. The idea goes something like this: when we talk about velocity, we are talking about an object traveling a certain distance in a certain amount of time. Thus, when we deal with velocities, we are actually dealing with the fundamental notions of space and time. What happens to those notions, he asked, if we assume that both Newton and Maxwell got it right?

Scientific folklore has it that he came to relativity one evening when he was going home from the patent office in a streetcar. He looked at a clock on a tower and realized that if his streetcar were traveling at the speed of light, it would appear to him that the clock had stopped. Time, in other words, seems to depend on the frame of reference of the observer. Since the principle of relativity tells us that all observers are, in some sense, equal, this means that there can be no universal time—no "God's eye" frame of reference in which time is "correct." This is, perhaps, the most fundamental difference between Einstein and Newton.

Working out the consequences of the principle of special relativity is fairly simple, involving little more than high school algebra and the Pythagorean theorem. The results are not what our Newtonian intuition tells us they should be, but have been abundantly verified by experiment in the years since 1905. They are:

Moving clocks run slower than identical stationary ones.
Moving objects shrink in the direction of motion.
Moving objects are more massive than identical stationary ones.
$E = mc^2$

The first reaction to these results is often to say that while Einstein's clock may appear to stop, in fact it is "really" keeping time in the usual way. This is an understandable point of view, but it is a profound violation of the principle of relativity. In effect, it says that there is a particular frame of reference in which the time is "right," a privileged frame of reference in which the laws of nature are correct. By extension, this means that the laws derived in other frames are wrong, in direct contradiction of the principle, not to mention a century's worth of experiment.

As mentioned above, it took Einstein over a decade to extend the principle of relativity to all frames of reference, including those involving acceleration. The reason has to do with the complexity of the mathematics involved. Instead of high school algebra, it involved a field called differential geometry that had been developed only a few years before Einstein used it. In any case, the most important thing for us to understand about the general theory of relativity is that it remains the best description of the force of gravity available to us.

You can picture the way that Einstein approached gravity by imagining a rubber sheet stretched taut and marked in squared off co-ordinates. Now imagine dropping something heavy on that sheet—a bowling ball, for example—so that the weight distorts the sheet. If you now roll a marble across the distorted sheet, its path will be changed by that distortion.

In general relativity, the fabric of space-time, like the rubber sheet, is distorted by the presence of matter, and this distortion, in turn, alters the motion of matter in its vicinity. While Newton would describe the interaction between the Earth and the sun, for example, in terms of forces and the law of universal gravitation, Einstein would describe the same thing in terms of the distortion of the geometry of space and time. This way of looking at gravity has been very successful in describing exotic objects like black holes, but, as we shall see in Chapter 12, it is profoundly different from the way the other forces of nature are described in modern theories.

Quantum mechanics

The word "quantum" means "heap" or "bundle" in Latin, and "mechanics" is the old word for the science of motion. Thus, quantum mechanics is the branch of science devoted to the motion of things that come in bundles. By the end of the nineteenth century, it had become obvious that the world inside the atom would have to be described by such a theory. The electron was discovered, showing that the atom was not indivisible, as the Greeks had supposed, but had constituents. And this, of course, meant that the study of the structure of the atom could become a legitimate field of scientific study.

One of the most extraordinary figures who pioneered this new field was Marie Sklodowska Curie (1867–1934). Born in what is now Poland into a family whose circumstances had been reduced by their support of the Polish independence movement, she eventually traveled to Paris, where she worked her way through the Sorbonne as a tutor. She married one of her professors, Pierre Curie, and the two of them began studying the newly discovered phenomenon of radioactivity. As part of her doctoral research, Marie established the fact that radioactivity was associated with the properties of atoms and not, as some had argued, with the bulk properties of materials. In a sense, this result opened what later became the field of nuclear physics. She and Pierre then turned to the study of ores that contained radioactive materials (they actually coined the term "radioactivity") and isolated two new elements—polonium (named after her native country) and radium. For this they shared the Nobel Prize in physics in 1903. She was the first woman to be so honored.

Tragically, Pierre was killed in a traffic accident in 1906. Marie took over the laboratory herself, eventually becoming the first woman professor at the Sorbonne. In 1911 she received a second Nobel Prize (in chemistry) for her pioneering work on radium and its compounds. She is the only person to have received two Nobel Prizes in two different sciences. Later in life she, like Einstein, became an international celebrity and made two trips to America to raise money to support her research.

Curie's study of radioactivity and Thompson's discovery of the electron were a clear indication that the atom was not a featureless bowling ball, as John Dalton had imagined, but had an internal structure. In 1911 New Zealand-born physicist Ernest Rutherford (1871–1937) performed a series of experiments that elucidated this structure. As an aside, we should note that Rutherford is unique in that he did his most important work *after* he got the Nobel Prize (in 1908, he received the prize in chemistry for identifying one type of radiation). Working at the University of Manchester, he set up an experiment in which particles emitted in radioactive decays (think of them as tiny subatomic bullets) were directed against a gold foil where they could scatter off of the gold atoms. Most of the "bullets" went straight through or were scattered through small angles, but a small number—about one in a thousand—came out in the backward direction. The only way to explain this is to say that most of the mass of the atom is concentrated in a compact structure which Rutherford dubbed the nucleus, with electrons circling in orbit around it. This image of orbiting electrons is familiar, and has become something of a cultural icon.

The most astonishing thing about Rutherford's discovery is the emptiness of the atom. If, for example, the nucleus of a carbon atom were a bowling ball sitting in front of you, then the electrons would be six grains of sand scattered over the area of the county you are in. Everything else is empty space!

Thus, by the 1920s a clear experimental picture of the atom had been developed. In the meantime, important changes were taking place on the theoretical front. In 1900 the German physicist Max Planck (1858–1947), working on a complex problem involving the interaction of electromagnetic radiation and matter, found that the only way to solve the problem was to assume that atoms could absorb and emit radiation in discrete bundles that he called "quanta." An atom, in other words, could emit one unit of energy or two, but not 1.35 units. This assumption may have solved Planck's problem, but it contradicted the basic ideas of the Newtonian–Maxwellian view of the world. This was repugnant to Planck—indeed, he called his adoption of quanta an "act of despair." Nevertheless, his 1900 paper is generally taken to be the beginning of quantum mechanics.

In 1905, Albert Einstein extended Planck's idea by saying that not only could atoms emit and absorb radiation in quanta, but that the radiation itself came in little bundles (they're now called "photons"). He used this idea to explain the photoelectric effect, the emission of electrons from metals on which light is allowed to impinge. It is interesting that it was this paper, and not relativity, that garnered the Nobel Prize for Einstein in 1921.

Then, in 1913, the Danish physicist Niels Bohr (1885–1962) applied the idea of quantization to electron orbits and showed that many properties of atoms could be explained if those orbits could occur only at certain distances from the nuclei. (For the *cognoscenti*, I'll mention that Bohr assumed that the angular momentum of orbiting electrons was quantized.) This meant that unlike the solar system, where a planet can orbit the sun at any distance whatever provided that it has the right velocity, electrons are constrained to orbit only at certain distances from the nucleus, and cannot be anywhere in between those "allowed orbits." For this work Bohr received the Nobel Prize in 1922.

By the 1920s, then, it was becoming clear that everything in the subatomic world came in little bundles or, in the jargon of physics, was quantized. At this point there came upon the scene a rather extraordinary group of young physicists, mostly German, loosely organized around an institute founded by Bohr in Copenhagen. They created the science of quantum mechanics and changed forever the way that human beings look at their universe. In the interests of space, we will consider only two men from this group—Werner Heisenberg (1901–76) and Erwin Schrodinger (1887–1961).

Heisenberg was an ardent German patriot, a point to which we'll return later. His main contribution to quantum mechanics is known as the "Heisenberg Uncertainty Principle." It basically says that there are certain quantities—the position and velocity of a particle, for example—that cannot, in principle, be known simultaneously. The reason is simple: in the quantum world the only way to make an observation is to bounce one quantum object off of another, and this interaction will change the object. You cannot, in other words, observe a quantum object without changing it. This makes the quantum world fundamentally different from our familiar Newtonian world, where observation does not affect the object being observed; that table doesn't change just because you look at it, for example.

Here's an analogy that may help you understand the Uncertainty Principle: imagine that you want to find out if there is a car in a long tunnel, but the only tool at your disposal is another car you can send down the tunnel. You could send the probe car down and listen for a crash, which would certainly tell you that there was a car in the tunnel—if you were clever, you could probably even find out where it had been. What you cannot do, however, is to assume that the car in the tunnel is the same after you observe it as it was before. In essence, the uncertainty principle guarantees that the interior of the atom will not be described by the familiar Newtonian rules, but by something new.

When the Nazis came to power in Germany, Heisenberg was faced with a painful choice. He could either abandon his country or stay and work with a regime he despised (and whose propagandists referred to him as a "White Jew"). In the end, he chose to stay and head up the German nuclear project—an enterprise similar to America's Manhattan Project that produced the atomic bomb. He concentrated on building a nuclear reactor, however, and never produced a weapon.

In 1998 a play titled *Copenhagen* by Michael Frayn opened in London and again raised the issue of Heisenberg's wartime activities. The play centers around a meeting in 1941, when Heisenberg came to Nazi occupied Copenhagen to talk to Bohr about the possibility of building nuclear weapons. The two men apparently talked for half an hour, then parted angrily and never spoke to each other again despite the fact that they had had an almost father and son relationship in happier times. In the aftermath of the play, letters were released that cleared up the mystery about the conversation. Heisenberg was actually breaking the law by discussing what we would call classified information with Bohr, so he spoke in vague general terms. He really wanted guidance on the morality of working on nuclear weapons. Bohr didn't understand what his younger colleague was talking about, and an argument ensued. Bohr was later smuggled out of Denmark and joined the Manhattan Project.

Schrodinger was Austrian, part of the educated upper class of Vienna's Jewish community. In 1926, as a professor in Zurich, he published what is now known as Schrodinger's Equation. Along with a mathematically equivalent theory published by Heisenberg, this equation represented a radical new way of thinking about the behavior of matter. A particle like an electron, which in the Newtonian world we would think of as something like a baseball, was represented as a wave—think of a tidal wave moving over the ocean. The height of the wave at each point is related to the probability that a measurement would show the electron to be at that point. But—and here is the great mystery of quantum mechanics—the electron really can't be said to have a position at all until the measurement is made. This way of looking at the world, now called the "Copenhagen interpretation of quantum mechanics," has been well verified by experiment since the 1920s, even though it seems paradoxical to us. Schrodinger received the Nobel Prize in 1933 for his work.

He actually produced a graphical way of thinking about the essential weirdness of quantum mechanics in 1935, when he published a paradox that has come to be known as "Schrodinger's cat." Suppose, he said, that you have a cat in an enclosed box with a vial of prussic acid. Suppose further that there is one atom of a radioactive material with instruments that will release the acid and kill the cat if the nucleus decays and emits radiation. According to the Copenhagen interpretation, the nucleus is both intact and has decayed until it is measured. Does this mean, Schrodinger asked, that the cat is half dead and half alive until we open the box and look in? (We don't have time to go into the extensive literature on this problem, but I will just mention that, at least as far as I am concerned, the paradox has been resolved.)

As was the case for many Jewish scientists, the rise of the Nazi party in Germany created problems for Schrodinger. He was offered a position at Princeton but didn't take it—probably because, as one biographer put it delicately, "his desire to set up house with his wife and mistress may have caused problems." In 1940 he accepted an invitation to found an institute for Advanced Studies in Dublin, where he stayed for the rest of his career.

I can't close this discussion of the development of quantum mechanics without a comment. When I am asked, as I frequently am, why America has become so dominant in the sciences, I say that we owe it all to one man—Adolph Hitler. The flood of scientists who sought refuge from the Nazi scourge forever enriched our own universities and changed the course of the history of science.

Once quantum mechanics was developed several important things happened. In the area of basic research, the probing of the subatomic world began in earnest. Throughout the middle third of the twentieth century, physicists discovered a veritable zoo of particles that lived out their brief lives inside the nucleus of atoms. There were literally hundreds of them, and they were mistakenly labeled as "elementary" particles. In the 1960s, in a strange replay of the explanation of the periodic table of the elements, theoretical physicists realized that the apparent complexity of the "elementary" particles could be understood if they were not "elementary" at all, but different combinations of things more elementary still—things called "quarks." The development of these ideas will be covered in Chapter 12.

On the less abstract side, the development of quantum mechanics led to everything we associate with the modern information revolution, from computers to iPhones and beyond. All of these devices depend on a single device— the transistor—that was invented by three scientists working at Bell Labs in 1947. John Bardeen, Walter Brattain, and William Shockely built a working transistor—it was about the size of a golf ball—that became the basis for the entire modern electronics industry. They shared the Nobel Prize in 1956.

In essence, a transistor is a device for controlling electrical current. It is analogous to a valve on a pipe carrying water. Just as a small amount of energy applied to the valve can have a large effect on the flow of water in a pipe, a small amount of electrical current supplied to a transistor can have a large effect on the electrical current flowing through the circuit of which the transistor is part. In particular, it can turn that current on or off, making the transistor the ideal tool for dealing with digital information, which is usually represented by a string of 0s and 1s. Since 1947, the size of transistors (as well as the methods of fabricating them) has undergone a profound change. Today, manufacturers routinely put millions of transistors on a chip the size of a postage stamp. It is this miniaturization that has made the modern information society possible, but it is important to remember, next time you use a computer or a cell phone, that it all depends on the strange laws of quantum mechanics.

The physicists' war

This is what some people have called World War II because of the tremendous contribution that physical scientists made to the war effort on both sides. A few examples of this sort of thing are the development of radar, which helped the Royal Air Force win the Battle of Britain, the invention of the proximity fuse by scientists at MIT and, of course, the thing that most triggers the association of scientists with wartime projects, the development of the atomic bomb in the Manhattan Project.

I have to point out that the association of scientific and technological advances with military endeavors is nothing new. Galileo, for example, was a consultant to the Arsenal of Venice, and did some of his most important scientific work figuring out how to scale up warship design, and some of the earliest work on projectile motion had to do with working out the trajectory of cannonballs. Nevertheless, World War II marked a turning point in this association, leading eventually to the modern electronic army.

The road to the atom bomb is usually reckoned to have started with a letter from Albert Einstein and other prominent scientists to Franklin Roosevelt pointing out that the conversion of mass to energy via the equations of relativity opened the possibility of a new kind of weapon. In addition, based on recent publications, they warned that German scientists were well on the way to developing these kinds of devices. In response, the United States undertook a top secret research program, eventually dubbed the "Manhattan Project" to develop the atomic bomb. By the early 1940s, the project had grown into a huge enterprise, employing over 100,000 people (including the best nuclear physicists in the United States and Britain) and involving research at over 30 locations in the United States and Canada. The most famous of these laboratories was Los Alamos, located in a remote part of New Mexico. It was here that the atomic bomb was developed.

The first atomic explosion took place in the New Mexico desert on July 16, 1945—the so-called Trinity test. In August, 1945, atomic weapons were dropped on Hiroshima and Nagasaki, ending World War II and initiating the nuclear age. The debate over whether the decision to use these weapons was justified continues to this day. One side maintains that as horrific as the casualties were, they were tiny compared to those which would have occurred had an invasion of Japan been necessary. The other claims that Japan might have been ready to surrender, making the attacks unnecessary. For what it's worth, the personal opinion of the author is that most of the evidence favors the first argument.

World War II was also a pivotal event in the history of science, because after the demonstration of what physics could do, governments in major industrial countries began supporting scientific research in a massive way. Many of the recent advances we have talked about are the result of this influx of funding.

The expanding universe

In the early twentieth century there was a serious debate among astronomers about objects they were seeing through their telescopes. Called *nebulae* (Latin for "clouds") they appeared in the best instruments as fuzzy patches of light in the sky. The debate centered around the issue of whether or not the nebulae were part of the Milky Way or, in the marvelous phrase used at the time, other "island universes"—what we would call galaxies today.

The resolution of this issue required two things: (1) a better telescope, capable of discerning the detailed structure of the nebulae, and (2) a scientist capable of using it. Fortunately, both were available in the early part of the twentieth century. The better telescope, with a hundred-inch mirror, was being built on Mount Wilson, near Los Angeles, by the industrialist and philanthropist Alexander Carnegie. The scientist was a remarkable man by the name of Edwin Hubble (1889–1953).

Hubble grew up in the suburbs of Chicago, and eventually attended the University of Chicago, where he was both a star athlete and honor student. He spent two years at Oxford as a Rhodes Scholar and, after trying out a couple of jobs on his return to the United States, returned to the university to get his Ph.D. in astronomy. On the day after he completed his degree, he volunteered into the army and served in World War I, coming back as a major in the infantry. In 1919, military service over, he joined the staff at Mount Wilson and took on the problem of explaining the nebulae.

The point is that with the new telescope Hubble could pick out individual stars in the haze of the nebulae. This had important consequences, because in the late nineteenth century Harvard astronomer Henrietta Leavitt had devised a scheme for finding the distances to a certain class of variable stars. These are stars that brighten and dim regularly over a period of weeks or months. Leavitt was able to show that the longer it took a star to go through the brightening and dimming cycle, the greater was the star's luminosity—the amount of energy it was throwing into space. By comparing this number to the amount of energy we actually get from the star, we can determine how far away it is.

The first important discovery Hubble made involved finding Leavitt's variable stars in a few nebulae and determining that the stars (and hence the nebulae) were millions of light years away from Earth. They were, in fact, other "island universes."

But he made a much more important discovery. Astronomers had known for some time that the light emitted from atoms in nebulae was shifted toward the red (long wavelength) end of the spectrum. This is an example of a common phenomenon known as the Doppler effect, and basically indicates that the nebulae are moving away from us. What Hubble was able to do was to combine his knowledge of the distance to each nebula with the knowledge of how fast that nebula was moving away from us (data obtainable from the red shift) to come up with a simple but revolutionary result: We live in a universe that is constructed in such a way that the farther away from us a galaxy is, the faster it is receding. The universe, to put it bluntly, is expanding!

This simple fact has a number of important consequences. For one thing, it means that the universe began at a specific time in the past—13.7 billion years ago, by current measurement. For another, it means that when the universe was younger, it was more compressed, and therefore hotter, than it is now. This will become important when we discuss the coming together of cosmology and particle physics in Chapter 11. The standard term used to refer to both the initial event and the subsequent expansion and cooling is the "Big Bang."

Having made these points, we have to stress that the Big Bang was not like the explosion of an artillery shell, with fragments flying out through space, but an expansion of space itself. Here's an analogy: imagine that you are making raisin bread, and that you're using a special transparent dough. If you were standing on any raisin as the dough was rising, you would see the other raisins moving away from you, because the dough between you and each of them would be expanding. Furthermore, a raisin that started out being three times as far away from you as another would be moving away from you three times as fast, because there is three times as much dough between you and the farther raisin. If you imagine that the raisins are actually galaxies, you would see

116 Physical sciences in the twentieth century

exactly what Hubble saw—a universal expansion. And just as no raisin is actually moving through the dough in our analogy, the Hubble expansion involves the galaxies being carried along by the expansion of space itself, not moving through space.

The fact that the universe was hotter when it was younger means that at very early stages in the Hubble expansion collisions between the constituent parts were more violent. This, in turn, means that as we go backward in time matter becomes broken down into its most fundamental components—from molecules to atoms to particles to quarks. Thus we arrive at the astonishing conclusion that to study the biggest thing we know about—the universe—we have to study the smallest things we know about—the elementary particles. We will return to this idea in Chapter 11.

Chapter 10

Biology in the twentieth century

In Chapter 8 we talked about what we called the "cellular turn"—the shift of attention of biologists away from organisms to the cells from which those organisms are made. This inward trend continued into the twentieth century, this time to the study of the molecules that operate within those cells. We will call this the "molecular turn," and we will see that it has brought about a fundamental shift in the way we look at living systems.

Just as new discoveries in the physical sciences in the twentieth century changed the way that people live, we can expect the molecular turn in the biological sciences to do the same in the twenty-first. Because advances in the biological sciences came later than those in the physical sciences, we are now just at the beginning of the biotechnology revolution, in roughly an analogous situation that the information revolution was in the 1960s. What happened in the twentieth century was nothing less than the acquisition of knowledge about the way that living systems work at the most fundamental molecular level. As we shall see, the century closed with the White House announcement of the first reading of the human genome, the blueprint for creating a complete human being from a single fertilized egg—a monumental achievement.

Although it is clear that the coming biotech revolution will alter human life in ways we can't imagine now, we are already starting to see its effect in our everyday experience. The majority of corn, soybeans, and cotton plants grown in the United States, for example, is genetically engineered to make it resistant to insect pests and various commercial herbicides. As we shall see in the next chapter, however, the grand medical advances that everyone anticipated would follow the sequencing of the human genome have yet to be realized. It turns out that the processes of life are a lot more complicated than anyone thought they would be, so that achieving the goal of what is called "genetic medicine" still remains elusive. In this chapter, then, we will concentrate on the process by which scientists unlocked the secrets of life.

I suppose we can start our story back in the nineteenth century, in the lab of a Swiss chemist by the name of Johannes Friedrich Meischer (1844–95). Working with white blood cells, Meischer isolated a substance he called "nuclein" in 1871, with the name indicating that the molecule came from the nucleus of the cell. Not a lot of attention was paid to this new discovery—nuclein was just one molecule among many being discovered at the time. Meischer had some vague idea that nuclein might be involved in heredity, but scientists at the time thought that heredity was too complex to be governed by a single molecule. Today, we realize that "nuclein" was nothing less than DNA, the central molecules in all of Earth's living things.

DNA as the molecule of life

Thomas Hunt Morgan (1866–1945) was born into a prominent Kentucky family. His father had been diplomatic consul in Sicily and was involved in Garibaldi's unification of Italy. His great grandfather was none other than Francis Scott Key, the composer of the *Star Spangled Banner*. After receiving his doctorate at Johns Hopkins in 1890 and, after a stint at Bryn Mawr, Morgan joined the faculty at Columbia University.

An important event happened in 1900, an event we have already discussed in Chapter 8. This was the re-discovery of Mendel's work on inheritance, an event that started many scientists thinking about genetics at the molecular level. Oddly enough, given the way his career was going to turn out, Morgan at first resisted the notion that a single molecule could control the development of an organism from a single fertilized egg to an adult. In essence, Morgan's belief that biology had to be an experimental science made him hostile to vague philosophical discussions of Mendel's "genes." The resolution of his difficulties came from a most unexpected source.

Consider, if you will, the common fruit fly, *Drosophilia melanogaster*. Like Mendel before him, Morgan had to choose an unlikely model for his work on the genetics of inheritance. Like Mendel's pea plants, Morgan's fruit flies have a conveniently short generation time—anyone who has left fruit out until it began to rot knows that fruit flies appear in a matter of days. (In fact, the adult fly feeds on the mold on rotting fruit.) Furthermore, fruit flies are easy to maintain—give them a jar with few pieces of rotting banana and they will happily procreate. In fact, when you read accounts of visits to Morgan's lab at Columbia, the universal thing remarked upon is the smell of the bananas.

Starting with eye color (red or white), Morgan's team began tracing the characteristics of fruit flies from one generation to the next. He had the good fortune to attract two brilliant undergraduates to his lab. Alfred Sturtevant and Calvin Bridges would go on to become lifelong collaborators and colleagues. They found that certain traits seemed to go together down the lines of heredity—white eyes and fuzzy wings, for example. They also knew that a crucial event in inheritance is a process known as recombination, in which segments of DNA are exchanged between the mother's and father's chromosomes. One day Sturtevant, still an undergraduate, was talking to Morgan when he realized that the frequency with which traits go together from one generation to the next must be a measure of how close the genes for those traits are on the DNA.

Here's an analogy to help you understand Sturtevant's insight: suppose you have a road map of your state and tear it into pieces, with an eye toward inserting those pieces in another map. If two towns are close together, they are likely to be on the same piece, while if they are farther apart, they are less likely to do so. In the same way, Sturtevant reasoned, two genes close together on the DNA are more likely to stay together during recombination than two genes that are far apart.

With this insight, the group produced the first crude genetic maps of the fruit fly genome and established a principle truth of genetics:

Genes are arranged in a linear fashion along the DNA molecule.

In passing, we should note that when Morgan received the Nobel Prize in 1933, he shared the prize money with Bridges and Sturtevant so that the two could pay their children's college tuition.

The notion that it is DNA that carries heredity and not some other molecule was confirmed in a brilliant experiment in 1952. Alfred Hershey (1908–97) and his colleague Martha Cowles Chase (1927–2003) worked with a series of viruses known as bacteriophages ("bacterium eaters"). These viruses consist of some DNA surrounded by a protein membrane. They grew bacteria in two different media, one containing radioactive phosphorus (which is part of the DNA molecule) and the other in a medium containing radioactive sulfur (which is contained in the protein coat of the virus). When viruses were allowed to attack these two different groups of bacteria, two different groups of viruses were produced—one with its DNA marked, the other with its protein shell marked. When these were allowed to attack bacteria, the tracers showed that the DNA went into the body of the bacterium (where reproduction takes place) while the protein stayed outside. This showed that, contrary to what some scientists had argued, it was the DNA and not proteins that carried the hereditary information in cells. For this work, Hershey shared the Nobel Prize in 1969.

The most famous development involving DNA, of course, was the discovery of the double helix in 1953. But every great discovery is preceded by important, if less known, work. This is certainly the case for the double helix. Let us discuss a few of these developments here.

Between 1909 and 1929, the Russian-American biochemist Phoebus Levene (1869–1940), working at Columbia University, had identified the basic molecular building blocks that make up DNA (we'll discuss the actual structure of the molecule below). These building blocks are sugars containing five carbon atoms, a phosphate group (i.e. phosphorous atom surrounded by four oxygen atoms), and a series of four molecules known as bases. They are named adenine, guanine, cytosine, and thymine, but are usually denoted by the letters A, G, C, and T. These molecules are the "letters" in terms of which the code of life is written.

In the late 1940s, the Ukrainian-American biochemist Edwin Chargaff (1905–2002) found two important properties of the bases—results now known as Chargaff's rules. First, he showed that in the DNA for a given species, the amounts of T and A are the same and the amounts of C and G are the same, but that the amount of these two pairs is different—in humans, for example, A and T each make up about 30 percent of the DNA (for a total of 60 percent) while C and G make up about 20 percent each. The ratio of T/A to C/G varies from one species to the next, but the amounts of T and A are always the same, as are the amounts of C and G.

In addition, scientists using a technique called X-ray crystallography were beginning to make direct studies of the structure of the molecule itself. This technique involved shooting X-rays at crystallized bits of DNA and observing how those X-rays scatter from the atoms in the molecule. From the scattering pattern of the X-rays, information about how those atoms are arranged, and hence about the structure of the molecule, can be obtained. The main center for research on X-ray diffraction from DNA was at King's College, London, in the Medical Research Council's Biophysics unit. The primary researchers were Maurice Wilkins and Rosalind Franklin (see below).

The main protagonists in the DNA story are two young researchers at Cambridge University—James Watson, an American biochemist spending a year at the university, and Francis Crick, an English theoretical physicist. This odd couple used all the chemical and X-ray data available to them and actually built a model of the DNA molecule out of metal plates. The double helix structure they found has become a cultural icon since then.

The easiest way to imagine a DNA molecule is to think of a tall ladder. The sides of the ladder are made from alternating sugar and phosphate molecules (see above). It's the rungs that are really interesting, however. They are made from the joining of two of the bases named above, one base attached to each side of the ladder. It turns out that if you examine the bases, A and T have two points where chemical bonds can be formed, while C and G have three. This means that the rungs of the ladder can only be AT, TA, CG, or GC. As we shall see, it is the sequence of these letters down one side of the ladder that carries the genetic information that allows every living thing on the planet to exist.

Now that you have the ladder, mentally twist the top and bottom to produce the double helix. For the discovery of the structure of DNA, Watson, Crick, and Wilkins shared the Nobel Prize in 1962.

In the matter of Rosalind Franklin

There is one issue surrounding the discovery of the double helix that has left the realm of pure science and entered the realm of popular culture, and that is the role of Rosalind Franklin in the whole process. Unfortunately, she died of ovarian cancer in 1958, at the age of 37. Since Nobel Prizes are never awarded posthumously, this means that she could not have shared in the 1962 prize, but debate about the importance of her work continues to this day.

What does not appear to be in dispute is that Wilkins showed results of her work—the famous Photograph 51—to Watson and Crick. Experts say that the photograph clearly indicated a helical structure. There is nothing unusual about this event—scientists share preliminary results all the time. The point, however, is that Wilkins did this without Franklin's knowledge. According to Watson in his book *The Double Helix*, this followed an altercation between him and Franklin—two prickly characters—after he had proposed a collaboration. It is hard to believe, however, that seeing Franklin's work didn't play a role in the eventual unraveling of the structure of DNA.

History is full of scientists who came close to getting Nobel Prizes—one commentator counted 23 who might reasonably have expected to share the Prize for DNA structure—Edwin Chargaff would certainly be in this group. Would Franklin have shared the prize had she lived, or would she have been the twenty-fourth disappointed aspirant? We'll never know, but the unabashed prejudice she faced in the 1950s British academic establishment doesn't give us any reason to think that her work would have been recognized by her contemporaries. As was the case with Marie Curie (see Chapter 9), the scientific establishment was again slow to accept the notion that women could do outstanding science.

Unraveling DNA

As we have pointed out repeatedly in our discussion, no scientific idea can be accepted unless it is verified in the laboratory. The discovery of the double helix was followed by many experiments that backed up the Watson–Crick model. I will describe just one of these, carried out by scientists at Cal. Tech in 1957.

The experiment depends on the fact that when cells divide, the DNA molecule in the original cell has to be copied so that each daughter cell has its own full complement of DNA. We don't have space to describe this replication process in detail, but in essence the DNA "ladder" splits down the middle of the rungs, and each side then assembles the missing parts of the "ladder" from material in the cell. The result is two identical "ladders," one for each of the daughter cells.

In this experiment, bacteria were grown in an environment in which the only nitrogen available was the isotope N-15, which is slightly heavier than the normal isotope you're breathing right now (N-14). The heavy nitrogen was taken up by the cells as they replicated their DNA, until eventually all of their DNA contained the heavier isotope. The bacteria were then put into an environment containing only the normal (lighter) isotope of nitrogen and the amount of heavy nitrogen in the DNA of each succeeding generation was measured. They found that the first generation of bacteria had half of the nitrogen in their DNA of the heavy variety, the second generation one fourth, and so on. This, of course, is exactly what the Watson–Crick model predicted, and the result was added to the panoply of evidence supporting it.

Once the structure of DNA was known, the next project was to understand exactly how the molecule played its role in living systems. We shall see that in a sense this work is still going on, but here we will simply recount the discovery of the genetic code and the development of the ability to sequence the genomes of many living things (i.e. read all of the letters on their DNA). Before we do so, however, we will have to take a moment to sketch the basic mechanisms that allow living things to function.

In Chapter 8, we saw that life is based on chemistry. The chemical reactions in living cells involve large, complex molecules, and their interactions require the action of another kind of molecule known as an enzyme. An enzyme is a molecule that facilitates the interaction between other molecules, but does not take part in the interaction itself—think of it as a kind of molecular real estate broker who brings buyer and seller together, but doesn't actually buy the house himself.

In living cells, the molecules that act as enzymes are known as proteins. The proteins, in other words, run the chemistry that makes the cell function. Proteins are assembled from building blocks known as amino acids—think of a protein as a string of different kinds of amino acids, something like multicolored beads strung together to make a necklace. Somehow the information contained in the bases of the DNA molecule is translated into the sequence of amino acids in a protein, and the protein then acts as an enzyme to run a chemical reaction in a cell. Unraveling this process, which is the basic mechanism of life on our planet, is the prime goal of modern biology.

The first step in this unraveling process was the elucidation of what is called the genetic code—the precise statement that tells you that if you have a specific sequence of bases in the DNA you will get a particular amino acid in the protein being assembled. The crucial experiment in this process was carried out in 1961–62 at the National Institutes of Health in Bethesda, Maryland. Marshall Nirenberg (1927–2010) and his post-doctoral fellow J. Heinrich Matthaei (1929–) assembled an intermediate molecule in the protein production process (known as RNA) that would come from repeating appearances of the same base in DNA. They then put their artificial molecules into a series of media, each containing a different amino acid. They found that their artificial molecule bound to one and only one kind of amino acid, giving them the first piece of the genetic code puzzle. Nirenberg received the Nobel Prize for this work in 1968.

In rapid succession, scientists filled in the rest of the code. The basic result is this: three base pairs on DNA—what scientists call a *codon*—codes for one amino acid in a protein. A string of codons on the DNA, then, determines the sequence of amino acids in one protein which, in turn, acts as an enzyme to run one chemical reaction in a cell. The sequence of bases that codes for one protein is, in fact, nothing other than Mendel's "gene." With this discovery, as we pointed out earlier, the gene moved from being a mental construct to a real physical entity—a string of bases along a DNA molecule.

As the twentieth century progressed and the importance of DNA became more and more apparent, a movement began to grow in the biological sciences to map out the details of the molecules, particularly the DNA that appears in humans. In the mid 1980s, a group of senior biologists in the United States began to talk about a grand proposal that eventually came to be called the Human Genome Project. The idea was to "read" all three billion bases in human DNA, a process called "sequencing" the human genome.

It seems strange now, when sequencing has become such an important part of science, to note that at the beginning there was serious resistance to this idea among biologists. Their main concern was that it would turn their field, which up to then had consisted of small, independent research groups, into corporate "Big Science." In addition, the tools available for sequencing in the 1980s were pretty primitive and time consuming. As one young researcher (who has since become a prominent scientist in his own right) confided to the author at the time, "I don't want my life's work to be that I sequenced from base pair 100,000 to base pair 200,000 on chromosome 12."

Fortunately, as the 1990s progressed, the sequencing process became more and more automated, as we shall describe below. In 1995 the first genome of a simple bacterium—*Haemophlus influenzae*—was sequenced, and in 1998 the first genome of a multicelled organism—the flatworm *Caenorhabditis elegans*—was added to the list. Throughout the decade, our ability to sequence more and more complex genomes increased, until today there are literally hundreds of genome sequences known.

The main players in the sequencing game were Craig Venter, then at Celeron, a private corporation he founded, and Francis Collins, head of the genome project at the National Institutes of Health. The technique that eventually cracked the human genome, developed by Venter, is called the "shotgun." Here's how it works: a long stretch of DNA is broken up into short, manageable segments, each of which is fed into a separate automatic sequencing machine. The output of these machines is then fed into a bank of computers that put the bits and pieces together and reconstruct the original sequence of the entire molecule.

If you were going to use the shotgun technique to read a book, you would tear up several copies of the book, give each scrap to a different reader, then write a computer program that could take those scraps and work out what the book was.

Once you understand the shotgun technique, you can understand why scientists referred to the human genome sequencing procedure as "assembling" the genome. You will also understand why the entire sequencing process is often referred to as a "bioinformatic revolution," given the heavy dependence of the process on computing power.

In the symbolic year 2000, in a ceremony at the White House, Venter and Collins announced the "first assembly" of the Human Genome. Many observers felt that this was the final exemplar of the Socratic dictum "Know Thyself." What followed as the twenty-first century progressed will be dealt with in Chapter 12.

The Grand Synthesis and the new evolutionary biology

As the basic molecular mechanisms of life were being elucidated in the twentieth century, a parallel molecular development was taking place in evolutionary science. The first of these developments, usually referred to as the "Grand Synthesis" or the "Modern Evolutionary Synthesis" took place roughly from the 1920s to the 1940s. In essence, it was the incorporation of various branches of science into the Darwinian framework to produce our modern evolutionary theory. It can be argued that this incorporation process never really stopped and is still going on today, as we shall see below.

The basic problem with evolutionary theory in the early twentieth century was that, although Darwin's concept of natural selection had been pretty well accepted in the scientific community, many of the mechanisms by which it functioned remained unknown. For example, as we pointed out in Chapter 8, Darwin recognized that there are variations between members of a given population, but had nothing to say about the origin of those variations. The rediscovery of Mendelian genetics in 1900 had created a debate about the question of whether or not Mendelian genetics was consistent with the idea of natural selection. Many geneticists (including, at the beginning of his career, Thomas Hunt Morgan) believed that mutations of genes would produce discontinuous change in organisms, rather than the gradual change implied by Darwin.

These problems began to be resolved in the 1920s, when a group of scientists started to develop a field now known as population genetics. In 1930, the British biologist Ronald Fisher (1891–1962) showed through the use of mathematics that the chance of a mutation leading to improved fitness in an organism decreases as the magnitude of the mutation increases, thus showing that Mendelian genetics could be consistent with the kind of gradual change (associated with small changes in DNA) envisioned by Darwin. He also showed that the larger a population was, the more variation there would be in the genes, and hence the greater the chance that a variation leading to increased fitness would appear. At the same time, his fellow Englishman J.B.S. Haldane showed that the same mathematical techniques could be applied to real world situations, such as the famous change in color of pepper moths in industrial England, and that the pace of natural selection could actually be much faster than people had thought. (Actually, Haldane is best known for his witty aphorisms, such as his response when asked what biology teaches us about God. "He has an inordinate fondness for beetles" was the response.)

Ukrainian born Theodosius Dobzhansky (1900–75) actually worked with Thomas Hunt Morgan when he came to the United States on a fellowship in 1927 and followed Morgan to Cal. Tech. His primary contribution to the Grand Synthesis was the redefinition of the evolutionary process in genetic terms. In his 1937 book *Genetics and the Origin of Species*, he argued that natural selection should be defined in terms of the changes of the distribution of genes in a population. This, in essence, recast the entire evolutionary discussion, shifting attention from organisms to genes—as good an example as we can find of the "molecular turn." Like Haldane, Dobzhansky is best known today for an elegant turn of phrase. The saying is actually the title of a 1973 essay—"Nothing In Biology Makes Sense Except in the Light of Evolution"—and is often used by modern biologists, especially when confronting Creationism.

Once genetics and natural selection were melded together, the next fields to be brought into the folds were field biology (the study of actual organisms in nature) and paleontology (the study of the history of life, which in the 1940s meant the study of fossils). Theodosius Dobzhansky, for example, worked on the genetics of fruit flies from different parts of what was the Soviet Union to establish that the same molecular processes were at work in wild populations as in the laboratory. The German-American biologist Ernst Mayr (1904–2005) began his contribution by redefining the concept of a "species" in 1942. It wasn't just a group of organisms that looked alike, he argued, but a group of interbreeding organisms that could not breed with organisms outside the group. Once again, the focus was shifting from the gross structure of organisms to their DNA. Mayr also proposed a detailed method for the process of speciation—geographic isolation. Here's an example of how the process might work: two populations of different butterflies begin to colonize two different valleys. One valley is moist and green, the other dry and rocky. Natural selection will begin to act on the two populations differently—for example, selecting for green color in one and brown in the other. The DNA of the two populations would begin to diverge, and when the divergence became large enough, Mayr argued, you would have two different species.

Many paleontologists in the mid-twentieth century rejected natural selection because they saw the (very incomplete) fossil evidence available at the time showing linear progress toward modern species. The American paleontologist George Gaylord Simpson (1902–84) examined the evolutionary record of the horse in detail and, in 1944, showed that the actual fossil record, when examined closely in detail, showed no evidence of steady improvement. In fact, he showed that the best way to visualize the evolutionary history of any organism is to see it as a kind of bush, with all sorts of branchings and dead ends. This leads to the notion of the history of life as nature trying all sorts of blind alleys until something finally works—a picture profoundly in tune with Darwinian natural selection. One of Simpson's most famous quotes illustrates this idea: "Man is the result of a purposeless and natural process that did not have him in mind."

With all of these developments, evolutionary theory was brought into the orbit of modern molecular biology and statistical methods. The incorporation of DNA into the study of past life has had a revolutionary effect on the way the past is studied. The basic idea is that if we compare the DNA of two living organisms, the differences between their DNA will depend on how long it has been since they shared a common ancestor. The more time since the branching event, the more chances there are for mutations in the DNA of the two organisms and the more differences we will find. Using this technique, paleontologists now construct elaborate family trees of all sorts of organisms. Our present idea that chimpanzees are the closest evolutionary relatives of human beings, for example, arose when scientists showed that there were more differences between human and gorilla DNA than between humans and chimpanzees.

Late twentieth-century notions

The Selfish Gene

In 1976, British evolutionary theorist Richard Dawkins (1941–) published a book with this title. It was mainly concerned with changing the focus of evolutionary thinking from a concern with organisms to a concern with the genes in DNA. The argument is that what matters in natural selection is the propagation of genes from one generation to the next, and that organisms are simply the vehicle by which that propagation is brought about. In a sense J.B.S. Haldane's comment—"I would lay down my life for two brothers or eight cousins"—can be thought of as anticipating Dawkins' arguments.

Punctuated equilibrium

In 1972, paleontologists Stephen Jay Gould (1941–2002) and Niles Eldridge (1943–) published a re-examination of the fossil record. They argued that instead of the slow, gradual accumulation of changes implicit in the Darwinian approach, evolution actually took place through a process of occasional sudden changes interspersed with long periods of stability. They called their theory "punctuated equilibrium," and for a while there was a serious, sometimes rancorous debate over exactly how evolution took place (but not, I hasten to add, over *whether* it took place). One side (I don't remember which) even accused the other of being Marxists, for reasons that escape me now and weren't all that clear at the time, either.

In any case, the dispute was eventually resolved in the only way such disputes can be resolved in science—by a careful examination of the data. It turns out that the correct answer to the question "Does evolution proceed gradually or in sudden steps" is "yes." Examples of both processes (and just about everything in between) can be found in nature. Our current knowledge of the way DNA works is completely consistent with this fact. We know, for example, that there are some genes that, in essence, control many other genes. A mutation in one of these "control" genes will obviously produce a big change in an organism, while a mutation in an isolated gene can produce an incremental, gradual change of the type envisioned by Darwin.

Evo-devo

In a sense, the appearance of evolutionary development biology (evo-devo, in the slang of scientists) is a continuation of the Grand Synthesis. Evo-devo looks at the way various organisms develop in their embryonic stage and attempts to find the grand evolutionary patterns governed by the organisms DNA. In particular, it attempts to find common patterns of embryonic development—different processes governed by the same or similar genes—across a wide variety of organisms.

For example, on the "evo" side, we find amazingly similar structures in a wide variety of animals. Take your arm, for example. It consists of one long bone connected to two long bones, which in turn, are connected to a complex of small bones in the hand. This structure has been whimsically characterized by one author as "one bone, two bones, lotsa blobs." You can see this same structure in the wing of a bat, the fin of a porpoise, the leg of a horse, and in countless other animals. In fact, we can trace it back 375 million years to an organism named "Tiktaalik," which was the first animal to make the transition from water to land. (The name, incidentally, comes from a word meaning "large fresh water fish" in the language of the local Iniktituk people on Ellesmere Island, where Tiktaalik was discovered.) Thus the basic structure of your arms and legs has been preserved over millions of years of evolution.

The "devo" side of evo-devo can be illustrated by looking at your hand. We all start out as a spherically symmetric fertilized egg, yet your hand is decidedly asymmetrical—it has top and bottom, left and right, front and back. How does the system know how to incorporate this asymmetry into the structure? The basic fact seems to be that there are genes that produce molecules that diffuse out into the growing cells, and the activation of genes in those cells is determined by the concentration of the molecules. The whole process is controlled by a gene called "hedgehog" (scientists are notoriously bad at naming genes), and this gene is found in a wide variety of animals. Experimenters, for example, have used mouse hedgehog to grow something like a foot (instead of a fin) in skate. Many other genes seem to perform the same sort of function in different organisms.

So as was the case with the physical sciences, the biological sciences at the end of the twentieth century were quite different from the same sciences at the century's beginning. Newer, deeper truths had been discovered and newer, deeper questions, were being asked.

Chapter 11

The new internationalization of science

Science is, by its very nature, no respecter of national or cultural boundaries. Nature is universal, and so is the study of nature. As we have seen, the precursors of modern science appeared in many places around the world. The actual appearance of modern science in western Europe in the seventeenth century was quickly followed by its spread to other parts of the world. In Chapter 6, we saw how science traveled to the periphery of Europe, and in this chapter we will look at the way that science became a truly international endeavor, a part of world history.

One measure of the current internationalization of science is the number of research publications authored by groups of scientists from different countries. In a 2011 report, England's Royal Academy found that fully 35 percent of all papers in major international scientific journals were the result of this sort of collaboration, and that this represented a 10 percent increase over the previous 15 years. Nowhere is this growth of international collaboration more dramatic than in the United States, where the number of these sorts of papers increased from about 50,000 in 1996 to about 95,000 in 2008. Thus, when we look at the way that modern scientists work, the trend toward internationalization is quite clear. I will also remark in passing that this trend is apparent to most individual scientists, although few could give you the kind of numerical evidence produced by the Royal Society.

Although these sorts of historical sea changes are notoriously hard to date precisely, the turn of the nineteenth century is probably as good a time as any to name as the beginning of the new internationalization. We will see later that the early twentieth century was when non-European scientists began making significant contributions to world research. It was also a time when Asian students began to attend western universities in significant numbers—the School of Agriculture at the author's Alma Mater, the University of Illinois, for example, has a longstanding tradition of training Indian graduate students.

Perhaps no story exemplifies this trend better that the foundation of Tsinghua University in Beijing—a place now known as the "MIT of China."

After the Boxer Rebellion (1989–1901) the victorious European powers imposed harsh indemnity sanctions on the weak Chinese government. In 1909 President Theodore Roosevelt proposed that a significant portion of the American indemnity be returned to China to support Chinese students who wanted to study in the United States. As a result, Tsinghua College (as it was then) was founded in Beijing in 1911. The university has had a tumultuous history through World War II and the Cultural Revolution, but has survived to take its place among the great scientific institutions of the world.

Multinational science

At the end of World War II, the center of gravity of world science was in the United States. This state of affairs arose because the contributions of scientists to the war effort had convinced the nation's leaders that supporting scientific research was the way to build both the American economy and the political power. No longer would the support of science be left to the private philanthropists like Andrew Carnegie, who had shouldered the burden in the first half of the century. The shift to massive government support of science was inspired by a 1945 report titled *Science: The Endless Frontier* (written, interestingly enough, by Vannevar Bush (1890–1974), then head of the Carnegie Institution). Bush's report, which seems commonplace today because so much of it has been implemented over the past decades, argued that scientific research was the basis for economic growth, and that the federal government was the logical agent to promote that research. In 1950, largely at his urging, the National Science Foundation was created. This agency remains one of the main sources of funds for basic research in the United States.

But even as the method for funding science was changing, another trend was starting that would force individual governments to seek multinational support for large research projects. Put simply, science began to be too expensive for single governments to support without assistance.

We can take elementary particle physics, the study of the fundamental constituents of matter, as an example of this trend. This field of physics requires a machine known as an accelerator. In essence, an accelerator takes a beam of particles such as protons or electrons and speeds them up to a high velocity. The beam is then directed toward a target, and in the debris of the collisions physicists search for information about how the target is put together. The higher the energy of the beam, the faster the particles are moving and the deeper into the target they can penetrate. High energy, in other words, translates into more knowledge of the basic structure of matter.

The first accelerators, built back in the 1930s, were desktop affairs—you could hold them in your hand and they were relatively cheap. As the field progressed, the accelerators got bigger and more expensive. When the Fermi National Accelerator Laboratory was founded outside of Chicago in the 1970s, the cost of the machine, the largest in the world for many decades, was in the hundreds of millions of dollars—a lot of money in those days. This machine was housed in a circular tunnel a mile across and probably represents the largest such project undertaken by an individual nation. In 1993 the follow-on machine, the so-called Superconducting Super Collider, which was to be housed in a tunnel 54 miles around under the plains of Texas, was killed by Congress when its projected cost ballooned to more than six billion dollars. In effect, pursuing knowledge about the basic structure of matter had become so costly that even the world's largest economy couldn't support it. As we shall see below, the current largest machine, the so-called Large Hadron Collider (LHC) in Switzerland, is supported by contributions from all of the world's developed nations. As science gets big and costly, in other words, simple economic forces drive it to become multinational as well.

American ascendancy following World War II was quickly countered by events in Europe and other parts of the world. After all, Vannevar Bush's argument about the importance of scientific research in economic development didn't apply only to the United States—it applied to any industrialized (or industrializing) society. The Europeans were simply among the first, after the United States, to put Bush's ideas into practice. Since then of course, government support of research has spread to Japan and other Asian nations, and, now, to China. We will look at the foundation of the Tata Institute in India as an example of this kind of institutional response outside of the western world.

In 1954, some 12 European countries founded an institute whose English title was "European Council for Nuclear Research," known by its French acronym as CERN. (The story is that when the "Council" was changed to Organization the original acronym was retained, causing Werner Heisenberg to remark that it "would still be CERN even if the name is not.") The laboratory was established outside of Geneva, Switzerland, symbolically placed right on the Swiss–French border. The idea was that each of the member states would contribute to the laboratory, in return for which scientists from that state would have access to the lab's facilities. Today, with the addition of former communist countries, CERN's membership has swelled to 20, with several states such as Romania and Serbia in negotiations to join. In addition, there are six states (including the U.S.) with "observer" status, and dozens more (particularly Third World countries) whose scientists have access to CERN and participate in research there. Indeed, sitting at the lab's cafeteria terrace with a magnificent view of the Alps, you hear so many languages being spoken that you could almost believe you were at a miniature United Nations.

This last fact illustrates an important aspect of the way that the scientific community works. Although major research facilities are located in industrialized countries, there is a clear understanding that special efforts must be made to bring in scientists from the rest of the world. American and European universities, for example, routinely make room for Third World students in their graduate programs, and professional organizations make an effort to see that things like scientific journals are widely available. In addition, it is not uncommon for Third World scientists and others deemed to be outside the mainstream to be invited to major international meetings, just to keep channels of communication open. During the Cold War, for example, invitations to American meetings were often extended to what were then called Iron Curtain countries (although some of the beefy "scientists" with suspicious bulges under their coats did cause some comment).

Throughout the last part of the twentieth century CERN served as a major center for scientific research. Probably the most important development from the point of view of everyday life took place in 1989 when two physicists at CERN—Tim Berners-Lee and Robert Cailliau—were working on a project that would allow computers in various labs to exchange data with each other. Although this started out as an attempt to make their research more efficient, it is generally recognized as the beginning of the world wide web. They developed what is called the "hyper text transfer protocol"—the familiar "http" you see on web addresses. (We should note that although the terms "web" and "internet" have come to be used as synonyms, the two have different histories, with the internet having been developed in the United States with support from the National Science Foundation and the Department of Defense.)

The development of the web is an example of a process to which we will return below—the production of enormous social change from seemingly impractical basic research. Had you asked Berners-Lee in 1989 whether his "protocol" would allow you to stay in touch with your friends or make an airline reservation, he would have had no idea what you were talking about. Yet that is precisely where his invention led.

A similar development took place in India. In 1944 one of the scientists involved in the Indian Atomic Energy program, Homi Bhabha, approached the Tata Trust, a major philanthropic organization in India with the idea that they should fund an institute dedicated to research in fundamental science and mathematics. Since then the institute has been involved in nuclear research, computing (the first Indian digital computer was built there) and, in 2002, was designated as a university. Today the institute runs particle accelerators, radio telescopes, and balloon observatories scattered all over the subcontinent. Although it was funded with private money, most of its support these days comes from the Indian government.

The new internationalization and big science

As we intimated above, one of the driving forces behind internationalization of science is the sheer size of the kind of research tools needed to do forefront research. In this section we will look at two modern international collaborations that typify this trend – the Large Hadron Collider in Switzerland and

Project IceCube at the South Pole. We will then look at the life stories of a few non-western scientists (most of them Nobel laureates) to see how the international nature of science played out in a few individual cases. Finally, we will discuss what we will call the "research pipeline," the way the new scientific discoveries make their way into everyday life in the modern world, as another example of internationalization.

IceCube

This project, which we'll describe in more detail below, is designed to detect the most elusive particle in existence, a particle called the neutrino ("little neutral one.") As the name implies, the neutrino has no electrical charge, and it has almost no mass. Furthermore, neutrinos interact extremely weakly with other forms of matter—billions of them are going through your body as you read this without disturbing a single atom, for example. In fact, neutrinos can travel through a lead sheet light years thick without any interaction at all. This makes them very difficult to detect. The basic strategy is to build a detector with a large number of atoms in it and hope that once in a while a passing neutrino will jostle one of them.

Despite the difficulty in detection, scientists find that detecting those rare neutrino interactions yields an abundance of information. This is because neutrinos are routinely produced in nuclear reactions, and thus are markers of the most violent events in the universe. IceCube is an instrument designed to detect the high energy neutrinos created in those events.

Arguably the most dramatic astronomical instrument ever built, IceCube is nothing less than a cubic kilometer of instrumented ice located at the South Pole. The project involves some 220 scientists from 9 countries. In this system, the ice acts both as target for incoming neutrinos and the medium that records their interactions.

The obvious question that occurs to most people when they learn about IceCube is "Why the South Pole?" The basic point, made above, is that when a particle like a neutrino interacts so seldom with ordinary atoms, the only way to collect sufficient data is to put a lot of atoms into the detector, so that neutrinos will interact with a few of them. For the kind of neutrinos that signal cosmic violence, you would need a tank of water a kilometer on a side to get this job done. We would never build something this big, of course, but the South Pole sits on a layer of ice 2,800 meters thick. In addition, as remote as it is, airplanes fly regularly to South Pole Station, so transportation of scientists and equipment is manageable.

The technique for building IceCube is simple to describe. High pressure heated water is used to drill holes down into the ice—a 2,500-meter hole takes about 40 hours to complete. A cable with a string of detectors is then lowered into the hole and the ice is allowed to freeze around it. When a neutrino triggers a reaction with an atom in the ice, flashes of light are produced when the newly created particles interact with the surrounding ice. These flashes are recorded by the instruments on the cables. From this data, computers at the surface can reconstruct the properties of the original neutrino. They can, for example, work out its energy and the direction from which it came.

As of January 2011, all of the 86 instrument strings had been installed and are taking data. One amazing aspect of this setup is that IceCube can even occasionally detect neutrinos caused by incoming cosmic rays colliding with atoms in the atmosphere above the North Pole, neutrinos that actually travel through the entire Earth to be captured in IceCube's instruments! Scientists expect IceCube to record over a million high energy neutrino events over a period of about ten years. This will give us a massive database for analyzing the most violent events in the universe. But of course, the most important thing likely to come as we start analyzing cosmic neutrinos—the most important thing that comes when we open any new window on the universe—is the thing we don't expect.

One interesting point about IceCube is that it could probably have been built by a single industrialized nation—it's not that expensive. The international aspects of the project, then, arise from the innate tendency of science to bring together people with similar research interests regardless of nationality. The situation with the LHC, to which we now turn, is somewhat different.

The Large Hadron Collider (LHC)

Today CERN is best known in the scientific community as the future world center for high energy physics research. This is because of a machine known as the Large Hadron Collider, the world's highest energy accelerator. (A word of explanation: "hadron"—"strongly interacting one"—is the term physicists use to denote particles like the proton that exist within the nuclei of the atom.)

Arguably the most complex machine ever built, the LHC occupies a 17-mile (27-km) tunnel over 500 feet underground. In keeping with the international nature of CERN, the tunnel actually crosses the French–Swiss border, so that the particles cross from one country into another innumerable times during their sojourn in the machine. The actual heart of the machine is two pipes maintained at high vacuum through which particles circulate in opposite directions. At specific places around the tunnel, the beams are allowed to come together—this is why the machine is called a "collider"—and for the briefest instant all of the energy of the particles is dumped into a volume the size of a proton, producing temperatures that have not been seen since the first fraction of a second in the life of the universe. Huge detectors—the size of large buildings—surround the collision areas and detect whatever comes out of the nuclear maelstrom.

We will talk a little about what scientists hope to find at the LHC in the next chapter, but at this point we can simply comment on the international nature of the machine. In the first place, with a budget of over nine billion dollars, it is the most expensive scientific instrument ever built. It is well out-side the budget of most national research agencies. Over 10,000 scientists and engineers from over 100 countries participated in the design and construction of the machine—the United States, among other contributions, built one of the mammoth detectors. Thus, the LHC can be taken to be a model for how sci-entific research will be conducted in the future, where the cost and complexity of instruments can only be expected to grow.

In September 2008, the first proton beams circulated in the machine, but a few days later a massive electrical problem shut the machine down, requiring a year for repairs. Some people criticized the CERN staff for this turn of events, but the author was reminded of a comment by rocket scientist Werner von Braun—"If your machine works the first time you turn it on, it's over-designed." In any case, by November 2009, the machine was again in operation, and shortly thereafter proton collisions at half energy were recorded.

The research pipeline

Throughout this book we have referred to the intimate connection between science—our knowledge of the universe—and technology—our ability to use that knowledge to improve the human condition. As the development of the web illustrates, even the most impractical seeming ideas can have enormous social consequences. In this section we will look in a little more detail at the way that process works in modern technological societies. Our basic mental image is what we call the research pipeline—the stages of development that take an idea from an abstraction to a real product. We will divide this pipeline into three regions: basic research, applied research, and research and develop-ment (R&D). The boundaries between the regions are rather fuzzy, but they do mark three different kinds of scientific activity.

We can define basic research as research that is done primarily to discover truths about nature, without regard to possible future utility. Sometimes you can determine that someone is conducting basic research simply by looking at the subject being explored. A scientist studying a supernova in a distant galaxy, for example, is unlikely to have practical applications of his or her work in mind. Often, however, the boundary of what can be considered basic research is blurry. A chemist exploring the properties of a new material, for example, may be looking for a deeper understanding of the interaction between atoms, which would surely qualify as basic research, or might be looking for properties of the material that would make it commercially useful, which would not.

In general, basic research in the United States is carried out in universities and government laboratories. Following the philosophy of Vannevar Bush, this work is largely supported by the federal government through agencies like the National Science Foundation, the National Institutes of Health, and the Department of Energy.

As the name implies, the next step on the research pipeline—applied research—is research conducted with a definite commercial or technical goal in mind. As indicated above, the boundary between basic and applied research is not sharp. Scientists studying the spread of combustion in a mixture of gasses might be doing applied research if they were trying to design a better automobile engine and basic research if they were studying flares on the surface of the sun, for example. In the United States, applied research is supported in part by government agencies like the Department of Defense and NASA, and in part by private corporations.

The final stage in the pipeline is research and development, usually abbreviated as R&D. This is where the idea that began with basic research and developed to a possible utility in applied research is turned into a commercially viable product. It is at this stage that non scientific factors like cost and market demand begin to enter the picture. R&D in the United States is conducted primarily by private corporations, though it is often supported by government contracts.

Since World War II the primary goal of research management in the United States has been to keep the research pipeline full, with basic research turning up new ideas, applied research developing the ideas that look as if they might prove useful, and R& D turning the result into a useful product. The model has worked quite well, and has been adopted by countries around the world, particularly in Europe and Asia. It has become, in short, another aspect of the new internationalization of science.

Some individual lives in science

The internationalization of science has been growing for some time, as we can see by a brief look at the lives of a few non-western scientists. We will begin by examining the work of Hideki Yukawa (1907–87) in Japan, which culminated in his being awarded the Nobel Prize in 1949, and then move on to look at the contributions of the Indian scientists Chandrasekhara Raman (1888–1970), who received the Nobel Prize in 1930, Satyendra Bose (1894–1974), whose work has figured in recent Nobel Prizes, though he himself never received the award and, finally, Subrahmanyan Chandrasekar (1910–95), who received the Nobel Prize in 1983. Since he began his life in India and ending it in Chicago, Chandrasekar is as good a symbol as any of the new internationalization of science.

Hideki Yukawa was born in Tokyo and educated at what was then called Kyoto Imperial University. In 1935, while a lecturer at Osaka University, he published a paper titled "On the Interactions of Elementary Particles" that completely changed the way we think about forces. Although he pioneered techniques in what is now called quantum field theory, his contribution is most easily understood in terms of the Uncertainty Principle.

According to Heisenberg, it is impossible to know both the energy of a system and the time at which it has that energy simultaneously. This means, in essence, that the energy of a system can fluctuate over a short period of time without our being able to detect the fluctuation. If the energy fluctuates by an amount $E = mc^2$, where m is the mass of a particle, then that particle can be present during that short time period without our seeing a violation of the conservation of energy. These so-called "virtual particles" are a mainstay of modern physics, but Yukawa introduced them to explain the force that holds the nucleus together—the so-called strong force. His idea was that the strong force could be generated by the exchange of an as yet undiscovered virtual particle between protons and neutrons in the nucleus. From the properties of the strong force, he predicted that there should be a particle about ⅐ as massive as the proton. When that particle was seen in cosmic ray experiments (it's now called the π meson or "pion"), the Yukawa theory of forces became a standard staple of particle physics. We will encounter the idea of virtual particle exchange again when we talk about the frontiers of modern science in Chapter 12.

C.V. Raman grew up in an academic household (his father was a lecturer in physics and mathematics) and, after obtaining his doctorate from Presidency College in what was then Madras (now Chennai), he joined the faculty to the University of Calcutta. It was there, in 1928, that he discovered a process now known as Raman scattering, which involves the interaction of light with atoms or molecules. When an atom absorbs light, it can use the energy to move electrons to higher Bohr orbits, and the electron normally falls back down to its original state, emitting light of the same frequency as it absorbed. Occasionally, however, the electron falls back down to a different orbit from the one in which it started, and the result is that light is emitted at a different frequency. This is Raman scattering, and is now widely used as a means of determining the chemical composition of all sorts of materials. Raman received many honors for his discovery besides the Nobel Prize—a knighthood, for example.

S.N. Bose was born in Calcutta, the son of a railroad engineer—the kind who designs systems, not the kind who runs a train. He taught at the University of Calcutta for several years before joining the faculty at what is now the University of Dhaka. What followed is one of the most bizarre episodes in the history of science. He gave a lecture on a particular problem in quantum mechanics, and made what seemed to physicists at the time to be a fundamental error in counting. Here's an example to show what I mean: suppose you were counting coin flips using two coins. If the coins were clearly distinguishable from each other—if one was red and the other blue, for example—then a situation in which you saw heads–tails would clearly be different from one in which you saw tails–heads. In Bose's argument, however, these two situations were seen as the same thing, something that could only happen if the two coins were indistinguishable. Despite this seemingly fundamental error, Bose's results seemed to match experimental results for particles like photons.

After several journals had rejected his paper, Bose sent it to Albert Einstein. Einstein immediately recognized the merit in Bose's work, personally translated it into German, and saw that it was published in a prominent German physics journal. As a result, the field of what is now called Bose–Einstein statistics was born and, as mentioned above, became the basis for a lot of modern physics research. The basic point is that many particles, like photons, are indistinguishable from each other, and cannot be treated like the red and blue coins in the above example. Bose's "mistake" turned out to apply to this entire class of subatomic objects. After a two-year stint as a researcher in Europe, Bose returned to the University of Dhaka, where he remained for the rest of his career.

S. Chandrasekar, or "Chandra," as he was universally known in the scientific community, was born in Lahore, in what is now Pakistan. He came from a professional family—his father was deputy auditor-general in a major railway company—and had to contend with family expectations that he would train for the civil service. His uncle, however, was C.V. Raman (see above), who had faced similar pressures and who provided a role model for an academic career. After a successful career as a student at Presidency College, he was awarded a fellowship to study in England. On the long sea voyage in 1930, the 19-year-old Chandra produced a paper that would, eventually, lead to his Nobel Prize.

The work involved a type of star called a "white dwarf." This is a star that has run out of fuel and collapsed under the force of gravity to a body about the size of the Earth. What keeps it from collapsing further is the fact that the electrons in the star can't be jammed together any further. For reference, the sun will become a white dwarf in about five-and-a-half billion years.

Preliminary calculations on the properties of these stars by the Cambridge scientists Chandra was going to work with had been made under the assumption that the electrons in the star were moving slowly. Chandra realized, however, that at the high temperatures in the stellar interior the electrons would be moving at an appreciable fraction of the speed of light, and this, in turn, meant that they had to be treated with Einstein's theory of relativity. This turned out to change everything, because his calculations showed that there was a maximum size a white dwarf can have, a size now known as the Chandrasekar limit (it's about 1.5 times the mass of the sun). Although the scientific community was slow to accept his results, it has become a mainstay of modern astrophysics.

In any case, he received his degree from Cambridge in 1933 and in 1936, while on a ship back to England, received an offer of a position at the University of Chicago. After marrying his fiancée, he moved to Chicago, where he stayed for the rest of his career, making contributions to many fields of theoretical physics. After his death the X-ray satellite now in orbit was named the "Chandra" in his honor.

Physicist Freeman Dyson has pointed out that Chandra's work on white dwarves marked a major shift in our view of the universe, breaking forever the hold of Aristotle on the modern mind. In the Aristotelian view, the heavens were the abode of harmony and peace, eternal and unchanging. What Chandra showed was that the universe is, in fact, a violent place, full of change and disruption. Even a placid star like the sun will change into a white dwarf, and later workers showed that larger stars exploded in gigantic supernovae, producing weird objects like pulsars and black holes. Today, the concept of a violent universe is commonplace, but in Chandra's time it was not. He was truly a pioneer in the cosmos.

Every scientist of a certain age, including the author, has a Chandra story. I was a frequent visiting faculty member at the University of Chicago, and remember him as a dignified presence in the physics department, always dressed in a gray suit, white shirt, and conservative tie. At a colloquium one day, a young physicist was presenting a talk on string theory (see Chapter 12) and brought up the point that progress in the field was slow because the mathematics was so difficult. At the end of the talk, Chandra got up and said that he and the others who had been developing general relativity in the 1930s had felt the same way. "Persevere," he said. "You will succeed just as we did."

What a marvelous thing to say to a young researcher!

Chapter 12

The frontiers of science

In this book we have traced the development of science from its primitive roots to its full blown modern form. We have, in other words, seen where science has been. It is fitting, then, to devote this last chapter to an exploration of where it is going. In fact, we can identify several important trends that have appeared in scientific work over the past half century—trends that can help put specific examples of scientific frontiers in context.

In Chapter 11 we identified internationalization as a growing trend in the sciences and discussed the rising cost of forefront research. It is simply a fact that as a particular science matures, the kinds of questions that get asked and the kind of equipment needed to answer those questions changes. I like to think that scientific research is analogous to the exploration of the American west by European settlers. In the beginning, when no one knew much about the topography of the land, the mode of exploration was something like the Lewis and Clark expedition—a quick survey to map out major features like mountains and rivers. Eventually, however, these initial exploratory expeditions were replaced by the Geological Survey—a painstaking mapping of the new territory, mile by mile.

In just the same way, when a new field of science opens up the first task is to map out its main features. This operation does not usually involve expensive equipment. Once that initial exploration is over, however, the task of working out the details begin. The questions become narrower and the investigations become more complex. This usually involves not only the work of more scientists but, as we saw in the case of elementary particle physics, more expensive equipment. This is true even for individual laboratories in universities, but when the expense gets high enough, it becomes a driving factor in promoting internationalization, as we have seen. Thus, we can expect this process of complexification to continue in the future, and to be a defining characteristic of twenty-first century science.

In what follows we will examine the problems that occupy scientists in a few fields today. This is not a comprehensive view of the frontiers of science, but rather a sample of how the trend toward complexification is playing out in a few fields.

String theory and the Higgs

The quest for an understanding of the basic structure of matter extends, as we have seen, back to the Greeks. We have seen how this understanding has deepened—how we learned that materials are made from atoms, how the nuclei of atoms are made of particles which, in turn, are made from quarks. The next (and possibly last) step in this process is the development of what are called string theories. In essence, these theories treat the ultimate constituents of matter (such as quarks) as being made of microscopic vibrating strings, with the main complication coming from the fact that the strings are vibrating in 10 or 11 dimensions. This makes the mathematics of the theories quite complex— a fact alluded to in the "Chandra" story in the last chapter. Nonetheless, some basic concepts of these theories will be tested at the LHC.

Most physicists would say that the first task of the LHC is to find a particle called the "Higgs." Named for Scottish theoretical physicist Peter Higgs who first proposed its existence, it is supposed to explain why particles (and, by extension, you and me) have mass.

A word of explanation: there are certain primary properties that material objects possess—electrical charge and mass are two examples. When physicists in the nineteenth century began examining the nature of electrical charge, enormous practical benefits flowed from their work—the electrical generators that provide power to your home, for example. Mass has always been a little bit more mysterious, with no one really being able to explain why any material object has it. This is why Higgs' theoretical predictions were so important, because if his particle is found at the LHC physicists will have, for the first time, a plausible explanation for mass.

Here's an analogy to help you understand how the Higgs mechanism works. Suppose you have two people in an airport concourse: a large, burly man dragging two large suitcases and a slim and unencumbered young woman. For the sake of argument, assume that both move along the concourse at the same speed when it is empty.

Now imagine what happens when the concourse is crowded. The man with the suitcases is going to be bumping into people, maneuvering around obstacles, and generally moving slowly. The young woman, on the other hand, will move through the crowd quickly. If the crowd in the concourse was invisible to us and we could see only our two protagonists, we might conclude that the man had a greater mass because he was moving more slowly. In just the same way, Higgs argued, we are surrounded by a sea of as yet unseen particles (particles that now bear his name) and the resistance to acceleration we normally associate with mass is actually nothing more than the interaction of ordinary matter with those particles. In our analogy, on other words, the unseen crowd plays the role of the Higgs particle. Theorists predict that collisions in the LHC will produce a few Higgs particles per month, so it may be a while before enough data accumulates to allow scientists to claim a discovery.

In addition, some versions of string theory predict the existence of a suite of as yet unknown particles. They are called "supersymmetric partners" of the particles we know about, and are usually denoted with an "s" in front of the particle name. The partner of the electron, for example, is called the selectron. The search for these particles is also a part of the LHC agenda.

The island of stability

When John Dalton introduced the modern atomic theory in 1808, chemists knew about a few dozen chemical elements. By the time Dmitri Mendeleev put together the first version of the periodic table of the elements in 1869, on the other hand, most of the naturally occurring chemical elements were known. The general situation that has come out of studying these elements is this: we know that the nuclei of the atoms are made from combinations of positively charged protons and electrically neutral neutrons. For light nuclei, the number of protons and neutrons are equal, but neutrons tend to predominate for heavier nuclei. The more protons there are, the stronger the electrical repulsion and the more difficult it is to make the nucleus stay together. Thus, the heavier the nucleus, the more likely it is that it will undergo radioactive decay unless more neutrons are packed in to "dilute" the electrical repulsion. Nature herself can do this up to uranium, which normally has 92 protons and 146 neutrons. This most common form of uranium undergoes radioactive decay with a half life of about 4.5 billion years—approximately the same time as the age of the Earth—so you can think of uranium as being almost, but not quite, stable. All nuclei past uranium (i.e. all nuclei with more than 92 protons) are unstable.

This has not prevented scientists from trying to produce and study the so-called "trans uranium" elements. In fact, since the 1940s, the production of these nuclei has been something of a cottage industry in physics. The basic procedure is to smash two lighter nuclei together at high energy and hope that, occasionally, the protons and neutrons in the resulting soup will arrange themselves in such a way as to produce an as yet unknown chemical element. These new elements tend to be very unstable, often decaying in less than a second.

As the decades went on, then, the elements beyond uranium began to be filled in. The names later given to these elements tell the story of discovery— Berkelium (element 97), Dubnium (element 105, named for the Joint Institute for Nuclear Research at Dubna, in what was then the Soviet Union), Darmstatium (element 110). Other elements were named for famous scientists—Einsteinium (element 99) and Mendelevium (element 101), for example. The basic trick has always been to try to cram as many extra neutrons into the new nucleus as possible, using those neutral particles to "dilute" the electrical repulsion of the protons, as discussed above. In April of 2010, scientists announced that they had smashed a calcium nucleus into a nucleus of berkelium to produce element 117, thereby filling in a blank in the periodic table, which is now completely filled up to 118.

None of these superheavy elements (a term generally used to describe nuclei beyond 110 or so) has been produced in large enough quantities to have any practical use, but that may change in the future. Scientists have known for a long time that nuclei that have certain numbers of neutrons and protons are unusually stable and tightly bound. These are sometimes referred to as "magic numbers" of protons and neutrons in informal discussions. Theorists predict that somewhere around element 126, these magic numbers may combine to produce elements that are once again stable. They refer to this as yet unreached region of the periodic table as the "island of stability." As they approach the "island," theorists expect that the lifetimes of their new nuclei will increase slowly, from milliseconds to seconds to hours to days to (perhaps) years. Some hope for this outcome came from the data on element 117, some forms of which lasted over 100 milliseconds.

To reach the island, physicists will need a machine capable of accelerating intense beams of very heavy nuclei, so that extra neutrons can be crammed into the resulting superheavies. Michigan State University is in the process of building a machine called the Facility for Rare Isotope Beams that will probably become the forefront machine for this type of research in the future.

Up to this point the production of superheavy elements has been of interest primarily to those trying to understand the structure of the atomic nucleus. The prospect of creating stable superheavy nuclei raises the possibility of discovering an entirely new kind of chemistry associated with these new atoms. Where such a development might lead, both in terms of basic research and potential applications, is anyone's guess.

Dark energy

In Chapter 9 we met Edwin Hubble and described his discovery of the expansion of the universe. If you think abut the future of the expansion, you realize that there ought to be only one force that can affect it—the inward pull of gravity. A distant galaxy, speeding away from us, ought to slow down because of the gravitational attraction of all of the other galaxies pulling it back. In the late twentieth century, measuring the so-called "deceleration parameter" was a major goal of astronomers.

The problem is that when the distance to a galaxy is so great that it just looks like a point of light, all of our standard methods of measuring distance fail, so it is impossible to know how far away the galaxy is. A new standard candle (see Chapter 9) is needed. Two teams—one at the University of California at Berkeley, the other at the University of Michigan—came up with just that in an event called a Type Ia supernova. These events occur in double star systems in which one of the partners has run through its life cycle and become a white dwarf (see Chapter 11). The white dwarf pulls material from its partner, building a layer of hydrogen on its surface that eventually detonates in a titanic nuclear explosion that destroys the star. Since the Chandrasekar limit defines the size of a white dwarf, we would expect all Type Ia supernovae to emit the same amount of energy. If we compare this energy to the amount we actually receive from a supernova in a distant galaxy, we can tell the distance of the galaxy in which the explosion occurred.

The key point to recognize here is that when we look at a galaxy billions of light years away, we are seeing the universe as it was billions of years ago, not as it is now. By measuring the red shift of those distant galaxies we can compare the rate of expansion of the universe then to what it is now. It was expected, of course, that the expansion would be slowing down under the influence of gravity. One of the biggest shocks the scientific community ever received came when these teams announced their results—the expansion wasn't slowing down at all, it was speeding up!

Obviously, there was some previously unsuspected force in the universe acting as a kind of anti gravity, pushing galaxies farther apart when their speeds should have been slowing down. Cosmologist Michael Turner of the University of Chicago called this mysterious substance "dark energy." Subsequent measurements have told us several things about dark energy. For one thing, it makes up about ¾ of the mass of the universe. For another, we can now trace out the expansion history of the universe. For the first five billion years or so, when galaxies were close together, the expansion did indeed slow down as gravity overpowered the effects of dark energy. After that, however, as galaxies got farther apart and the attractive force of gravity weakened, dark energy took over and the current accelerated expansion began.

It is a sobering thought to realize that we have no idea at all about the stuff that makes up most of the universe. Elucidating the nature of dark energy is certainly the most important outstanding problem in cosmology, quite likely the most important problem in the physical sciences.

Genomic medicine

When the human genome was sequenced in 2000, commentators foresaw a massive improvement in medical treatments. After all, many diseases result from the failure of specific cells in the body to perform normally—diabetes, for example, involves the failure of cells in the pancreas to produce insulin. Since in the end the function of cells is controlled by DNA, our new understanding of our own genomic makeup should, they argued, revolutionize the practice of medicine.

On the one hand, if our genes really tell the story of our lives, reading an individual's genome should tell the physician what diseases and conditions are likely to develop, and appropriate measures could be taken. A person predisposed to heart problems, for example, might be told to follow an exercise and dietary regimen to lower his or her risk. On the other hand, once a cell has failed to function normally, a knowledge of the genome might lead us to new methods of treatment, such as replacing defective genes in the malfunctioning cell.

Perhaps the most ambitious scheme involving genomics goes by the name of "regenerative medicine." To understand it, we need to make a slight diversion and discuss the nature of stem cells.

Everyone reading this book began his or her existence as a single fertilized cell, with one molecule of DNA. It is the repeated division of that cell and that attendant copying of the DNA that eventually produce an adult human being. Thus, every cell in our bodies contains the same DNA, but as the fetus grows, most of the genes are turned off, with different genes operating in different cells. For the first few divisions, however, all genes are capable of operating, and it is only later that specialization "flips the switches" and starts turning genes off. A cell with all genes operational is called a stem cell.

We know that if we remove the nucleus that contains the DNA from an egg and replace it with the nucleus of an adult cell from another individual, the egg knows how to reset all the switches that have been thrown in the adult cell (although we don't know how this process works). Thus, it is possible to produce, in the laboratory, a stem cell containing the DNA of the individual donor of that DNA. The basic idea of regenerative medicine (and I have to stress that aside from producing stem cells none of this can be done right now) is to harvest the stem cells after a few divisions and use them to grow new tissue. Most likely the first application will be growing neurons for the treatment of Parkinson's disease, followed perhaps by pancreatic beta cells to produce insulin. The point is that the new tissue will have the DNA of the patient, and will therefore not be rejected by the body's immune system as happens with transplanted organs now. If this program is successful, you can imagine a world where worn out or diseased body parts are exchanged for newly grown ones, more or less in the way we replace parts in our cars today.

It is my impression that the social consequences of this sort of medicine have not been well thought out. For example, a full blown regime of regenerative medicine would greatly increase the lifespan of those who have access to it, a development that might have all sorts of unanticipated consequences. When I want to make his point to academic audiences I ask "Would you really want to be an assistant professor for five hundred years?"

Regenerative medicine would also make answering a simple question like "How old are you?" somewhat problematical, since the answer could well be different for different body parts. The birth rate would have to drop significantly if overpopulation is to be avoided, and we really haven't thought about what a society with very few children would look like. Finally, there would be the simple problem of boredom—how many times can you see *Hamlet*? We could go on, but I think the point has been made—when and if regenerative medicine becomes a reality, it will raise questions far beyond those involved in the science.

Looking over the first decade of the twenty-first century, however, one fact stands out: virtually none of the grand predictions made for genomic medicine have been realized. It is useful to ask why the revolution has been delayed.

The basic reason for this state of affairs is that it turns out that interpreting the genome is a lot more complicated than people thought it would be in 2000. For one thing, the hope that we would find one (or at most a small number) of genes implicated in specific diseases was not realized. Even though there are examples of this sort of connection—cystic fibrosis, for example, is associated with damage at a specific site on one chromosome—we are finding that most diseases are much more complex. If subtle relationships between hundreds or even thousands of genes have to be examined to deal with a disease, we are going to need to know a lot more about the genome than just the sequences of bases in the DNA. It may even be, as we shall discuss below, that we will need to develop new mathematical tools to deal with the complexity we are finding in the genome.

An additional source of unanticipated complexity has to do with the fact that the functioning of the genes in DNA is managed by a delicate interplay of control mechanisms. A new field of science, called "epigenetics" (literally "outside the genes") has grown up to examine these processes. The basic idea is that without changing the sequence of bases in a gene, the gene can be turned off (for example, by attaching another molecule to the DNA to inhibit the normal working of the gene). These kinds of changes seem to follow cells through the process of division. It is, in fact, precisely these sorts of epigenetic changes that produce the cell differentiation we saw when we talked about the process by which cells differentiate as an individual matures. Epigenetic processes thus introduce yet another layer of complexity into understanding the genome.

The rise of complexity

As we have seen, modern science developed slowly as scientists acquired the capability of dealing with more and more complex systems. And although we have not emphasized this point up to now, these changes often follow the introduction of new mathematical techniques. It wasn't until Isaac Newton and Gottfried Leibnitz invented the calculus, for example, that realistic calculations of orbits in the solar system could be carried out. The introduction of the digital computer in the mid-twentieth century is another of those mathematical milestones that has greatly expanded the natural systems amenable to scientific analysis. The reason is simple: a computer can keep track of thousands of different factors that affect a system, something that it would be difficult for an individual human to do.

As computers have improved in capability over the decades, the kinds of problems that scientists can solve have become more complex. To take an example from my personal experience, in the early 1980s engineers designing airplanes did not have the ability to calculate the flow of air over an aircraft wing as the plane went through the sound barrier. There were just too many variables, too many things for the computers of the day to keep track of. Today a calculation like that could probably be done on a standard laptop, maybe even on a cell phone. This sort of progress in mathematical capability has given rise to a whole new field of science—the field of mathematical modeling. It is not too much to claim that the standard description of science as the interplay between theory and experiment must now be amended to include modeling as a third entry.

A computer model always starts with some basic scientific laws. If we want to make a model whose purpose is to predict the weather, for example, we would begin with the laws that govern the motion of atmospheric gasses, the laws that describe the flow of heat, the laws that describe humidity and condensation, and so on. At this level the model simply collects the relevant scientific knowledge available. If precise knowledge is not available (as is the case for the formation of clouds, for example) the known laws are supplemented by informed guesses as to what the as yet unknown laws might turn out to be. Once all this has been assembled, relevant data (for example, the temperatures at various weather stations) can be added.

At this point the computer is ready to go to work. The atmosphere is divided into boxes, and the working out of the laws in each box calculated. Does the temperature go up or down, does air flow in or out, etc.? Once this has been done for all the boxes, a grand accounting takes place—air leaving one box, for example, is added to neighboring boxes and so on. With the new arrangement, the computer is ready to re-calculate with the new parameters. Typically, the computer will step forward in twenty-minute intervals to get a prediction of the weather.

Computer models like this have become ubiquitous in science and engineering—we no longer test airplane designs in wind tunnels, for example, but "test" those designs using a computer model. As we look into the future, we can expect to see more of these kinds of complex systems modeled on computers. While this represents a major advance in science, it also carries risks. For example, much of the current debate on climate change centers around the results of complex climate models—models so complex that it is not an exaggeration to say that there is not a single person on the planet who understands everything in them. This raises problems when model results are presented, because there is always a nagging question of whether or not some obscure assumption might be wrong in a way that makes those results wrong or misleading. Presumably as time goes by and the models improve and, more importantly, we see that they make correct predictions, these kinds of doubts will diminish.

This admittedly cursory look at a few areas of current research illustrates an important point about science: so long as there is a universe out there to explore, there will be questions to be asked and investigations to be carried out, in fields as far apart as the study of the basic structure of matter, the design of the cosmos, or the working of the cells in our body. We will never reach the end of science, and each new twist and turn in the road leads to exciting new insights into the world we inhabit.

Revolution or continuity?

I distrust the word "revolution" when it is applied to intellectual history. Every event, no matter how novel, has precedents and antecedents, and thus can be viewed as part of a continuum. Thus, one way of looking at the development of science since the end of the nineteenth century is to see it as a kind of outgrowth and elaboration of what came before. Yet one can reach a point where "more" becomes "different," and I would argue that somewhere during the last century science crossed that boundary. The science of today is so different from what it was a century ago that it can truly be considered to be qualitatively different.

We can, in fact, identify two major factors that have produced this change. One is the massive influx of government funding that followed World War II, driven by the notion that keeping the research pipeline full is the best way to keep the economy growing. The other is the development of modern electronics, and particularly the digital computer, that have allowed the examination of nature at unprecedented levels of complexity. Historians in the future will doubtless point to some date in the last half of the twentieth century and say "This is where it all changed."

Epilogue

There is no question that the development of the scientific world view has had a profound effect on human beings, both in the practical and intellectual sense. Looking back over our story, we can see at least three different ways to think about the role and impact of science. One is intellectual, dealing with the way science has changed and transformed humanity's view of its own role in the universe. This way of looking at the history of science gets us into the fraught boundary between science and religion. A second point of view involves the way that scientific knowledge has changed the material conditions of human life. And finally, we can look at this history from the point of view of the internal requirements of science itself, viewing it as a purely intellectual exercise in the development of the scientific method. Let's consider each of these points of view briefly.

Science and religion

Human societies started out in what Carl Sagan called the "Demon Haunted World," an unpredictable world governed by the capricious whims of the gods. As scientific knowledge increased, the boundary between science and religion changed. More of the world was explained in terms of impersonal natural laws, less in terms of the will of the Gods.

This change was not always peaceful—witness the trial of Galileo—but by the twentieth century a stable *modus vivendi* had developed between the two areas of thought. The basic idea is that science is the way to answer certain kinds of questions about the natural world, while religion answers different questions about a different human reality. The kinds of questions asked in the two areas are different, as are the methods for answering them. To use an infelicitous but useful phrase coined by Steven Jay Gould, they are "non-overlapping magisteria." In less formal language, we can capture this difference by noting that science is about the age of rocks, religion about the Rock of Ages.

Looked at in this way, there is no cause for conflict between science and religion because they deal with completely disjoint areas of existence. Conflict can arise only when those in one area try to impose their views and methods on the other, a process we can call disciplinary trespass. Today, we can identify two such areas of trespass: one is the attempt by Creationists to impose religious beliefs in the teaching of evolution, the other is the attempt of scientific atheists to push the boundary the other way and provide "scientific" proofs of the non existence of God. Both attempts, I suspect, are doomed to failure.

Humanity's place in the universe

We started out with the notion that the Earth, the home of humanity, was the unmoved center of the universe and that humanity had a place in creation different from other animals. This comfortable view has not been treated kindly by the advance of science. First, the Copernican Revolution established that the Earth was just one planet circling the sun, then we learned that the sun is just one star in a galaxy full of stars, and finally, Edwin Hubble taught us that our galaxy is just one among a myriad of galaxies.

While this demotion of the Earth was in progress, Charles Darwin taught us that human beings were not all that different from other life forms on our planet. In fact, twentieth-century science has clearly established the fact that humans are intimately connected to other living things on Earth at the biochemical level.

So if we don't live on a special planet, and if our chemistry isn't all that different from that of other living things, is there anything that might make human beings special? Actually, the answer to this question can be found in the discussion of complexity in Chapter 8. As it happens, the human brain may be the most complex, interconnected system in nature. It could be that phenomena like consciousness that arise from that complexity are rare in the universe, in which case there might be something special and unique about human beings and their planet after all.

Material well being

This is probably the easiest aspect of the social implications of science to discuss, since it is the subject most people think of when they think of the benefits of science. From the invention of fire to the invention of the microchip, human understanding of the workings of nature has been used to better the human condition. Traditionally, major advances in the past—think of the steam engine and the electric light—have come from what we can call tinkerers rather than scientists. In the past it wasn't essential to understand the laws of nature that made something work to produce something useful—indeed, it can be argued that the invention of the steam engine triggered the development of the science of thermodynamics, rather than vice versa.

All that changed in the mid-twentieth century—you can take Vannevar Bush as a symbol of this development. Today useful devices are developed in the research pipeline discussed in Chapter 11—the pipeline that begins with basic research and ends on the shelves of your local supermarket. This is a profound (and often unremarked) change in the way that technology affects our lives.

The products of science and technology have not been an unalloyed benefit, however. For example, when Alfred Nobel invented dynamite, he made possible massive construction projects like America's Interstate Highway System, but he also made war much more destructive. The ability to tap the energy of the nucleus allowed us to generate a significant fraction of our electricity without contributing to global warming, but it also produced nuclear weapons. Once scientific discoveries are made they pass from the control of scientists to society at large. Furthermore, it is simply a fact that it is usually impossible to predict what a particular line of research will produce in the way of practical applications. Short of stopping research completely—a practical impossibility in today's world—we just have to be aware of the two-edged nature of scientific advance and deal with each new situation as it arises.

The scientific journey

In this book we have traced the development of science from its earliest appearance in the astronomy of Stone Age people to its present form in modern technological societies. It has been a long road, but there are several milestones that were significant.

The first of these was the dawning realization that the world is regular and predictable—that we can observe the world today and expect that much of what you see will still be there tomorrow. We symbolized this development by talking about structures like Stonehenge in England and the Medicine Wheel in Wyoming. This is the basis on which all later science is built.

The next important development took place in ancient Greece, when philosophers began to think about the world in what we would call naturalistic terms, explaining what they saw through the operation of impersonal (and predictable) laws rather than the unpredictable whims of the Gods. With this development, humanity began to leave the "Demon Haunted World" we referred to above. Much of this knowledge of antiquity was preserved and expanded by Muslim scholars during the first millennium CE.

The birth of modern science spanned several centuries in Europe, and culminated with the work of Isaac Newton in seventeenth-century England. For the first time, the full blown scientific method was available for the investigation of nature, and rapid advances in physics, chemistry, and biology followed. By the end of the nineteenth century the basics of the scientific world view were in place. We understood that the Earth was not the center of the universe and that human beings were connected to the rest of life on our planet by ties of descent and biochemistry.

The twentieth century has seen an enormous expansion of the world available to science. We have discovered that we live in a universe of billions of galaxies, and that that universe is expanding. We have also discovered that the kinds of materials we have studied throughout history make up less than 5 percent of what the universe contains—our ignorance is truly astounding. While we were exploring the large scale structure of the universe, we were probing down into the smallest things we can imagine, from atoms to nuclei to elementary particles to quarks to (perhaps) strings. At the same time we have learned the basic chemical processes that run living systems, and are in the process of unraveling the complexities of DNA.

No one can really predict where these new threads of research will lead, or what the world will look like a century from now. The story of science we have been following, then, should really be thought of as a kind of prologue to the unimaginable developments that are to come.

Index